Numerical Recipes
Example Book (C)

Numerical Recipes
Example Book (C)

William T. Vetterling

Polaroid Corporation

Saul A. Teukolsky

Department of Physics, Cornell University

William H. Press

Harvard-Smithsonian Center for Astrophysics

Brian P. Flannery

EXXON Research and Engineering Company

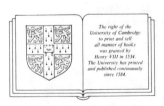

The right of the
University of Cambridge
to print and sell
all manner of books
was granted by
Henry VIII in 1534.
The University has printed
and published continuously
since 1584.

Cambridge University Press
Cambridge
New York   New Rochelle   Melbourne   Sydney

Published by the Press Syndicate of the University of Cambridge

The Pitt Building, Trumpington Street, Cambridge CB2 1RP

32 East 57th Street, New York, NY 10022 U.S.A.

10 Stamford Road, Oakleigh, Melbourne 3166, Australia

First published 1988

Reprinted 1988 (twice), 1989

Printed in the United States of America

Typeset in TEX

The computer programs in this book, and the procedures in *Numerical Recipes in C: The Art of Scientific Computing*, are available in several machine-readable formats. To purchase diskettes in IBM PC or Macintosh format, use the order form at the back of this book, or write to Cambridge University Press, 510 North Avenue, New Rochelle, New York 10801. The main book, example books, and diskettes are available in editions in the FORTRAN, Pascal, and C languages. Technical questions, corrections, and requests for information on other available formats and additional software products should be directed to Numerical Recipes Software, P.O. Box 243, Cambridge, Massachusetts 02238. Inquiries regarding mainframe and workstation licenses should also be directed to this address.

Library of Congress Cataloging-in-Publication Data

Numerical recipes example book (C).

1. C (Computer program language)

I. Vetterling, William T.

QA76.73.C15N86   1988   005.13′3   87-33843

ISBN 0-521-35746-2

ISBN 0-521-35467-6 (diskette)

CONTENTS

Preface . . . . *vii*

A Note on Numerical Recipes Utility Functions . . . . *ix*

1. Preliminaries *1*

2. Linear Algebraic Equations *5*

3. Interpolation and Extrapolation *22*

4. Integration of Functions *36*

5. Evaluation of Functions *44*

6. Special Functions *53*

7. Random Numbers *81*

8. Sorting *102*

9. Root Finding and Sets of Equations *113*

10. Minimization and Maximization of Functions *125*

11. Eigensystems *140*

12. Fourier Methods *150*

13. Statistical Description of Data *165*

14. Modeling of Data *187*

15. Ordinary Differential Equations *202*

16. Two-Point Boundary Value Problems *212*

17. Partial Differential Equations *218*

Appendix A   Header Files *223*

Appendix B   Numerical Recipes Utility Functions *229*

Appendix C   Functions for Complex Arithmetic *234*

Index of Numerical Recipes Demonstrated *237*

Preface

This *Numerical Recipes Example Book (C)* is designed to accompany the text and reference book *Numerical Recipes in C: The Art of Scientific Computing* by William H. Press, Brian P. Flannery, Saul A. Teukolsky, and William T. Vetterling (Cambridge University Press, 1988). In that volume, the algorithms and methods of scientific computation are developed in considerable detail, starting with basic mathematical analysis and working through to actual implementation in the form of C functions. The routines in *Numerical Recipes in C: The Art of Scientific Computing*, numbering over 200, are meant to be incorporated into user applications; they are functions, not stand-alone programs.

It often happens, when you want to incorporate somebody else's routine into your own application program, that you first want to see the routine demonstrated on a simple example. Prose descriptions of how to use a function (even those in *Numerical Recipes*) can occasionally be inexact. There is no substitute for an actual demonstration program that shows exactly how data are fed to a function, how the function is called, and how its results are unloaded and interpreted.

Another not unusual case occurs when you have, for one seemingly good purpose or another, modified the source code in a "foreign" function. In such circumstances, you might well want to test the modified function on an example known previously to have worked correctly, *before* letting it loose on your own data. There is the related case where function source code may have become corrupted, e.g., lost some lines or characters in transmission from one machine to another, and a simple revalidation test is desirable.

These are the needs addressed by this *Numerical Recipes Example Book (C)*. Divided into chapters identically with *Numerical Recipes in C: The Art of Scientific Computing*, this book contains C source programs that exercise and demonstrate all of the *Numerical Recipes* procedures and functions. Each program contains comments, and is prefaced by a short description of what it does, and of which *Numerical Recipes* routines it exercises. In cases where the demonstration programs require input data, that data is also printed in this book. In some cases, where the demonstration programs are not "self-validating," sample output is also shown.

Necessarily, in the interests of clarity, the *Numerical Recipes* functions are demonstrated in simple ways. A consequence is that the demonstration programs in this book do not usually test all possible regimes of input data, or even all lines of source code. The demonstration programs in this book were by no means the only validating tests that the *Numerical Recipes* functions were required to pass during their development. The programs in this book *were* used during the later stages of the production of *Numerical Recipes in C: The Art of Scientific Computing* to maintain integrity of the source code, and in this role were found to be invaluable.

A Note on Numerical Recipes Utility Functions

The programming conventions used in this book are discussed fully in Chapter 1 of *Numerical Recipes in C: The Art of Scientific Computing.* You should have a copy of that book – without it, this one will not be very meaningful. Nevertheless, we review a few important matters here.

The *Numerical Recipes* software collection in C contains over 200 function routines. For the most part the routines are self-contained, making reference only to other routines in the package, or to standard C library functions. There are some consistent exceptions, however: (1) When the *Numerical Recipes* functions, (or the main() programs in this book) encounter an error, they turn matters over to a function nrerror(), not a standard C function. (2) When vectors and multidimensional matrices are allocated or deallocated, use is made of a set of utility functions with names like matrix(), vector(), free_matrix(), and free_vector(). These Numerical Recipes-specific functions are found in a utility file nrutil.c, listed in Appendix B. Example routines that use any of these utilities include the header file "nrutil.h" (Appendix A) along with any of the standard library header files which are required. Similar remarks apply to routines using complex arithmetic. These make calls to functions in the file complex.c, which is listed in Appendix C. The appropriate header file is called "complex.h" (Appendix A).

All of the examples in this book call at least one, and sometimes more than one, of the *Numerical Recipes* functions. A type declaration must generally be given for each of the functions used. To spare readers the trouble of searching program listings for the proper declarations, we have compiled an alphabetical list in Appendix A. This list also comprises the contents of the header file nr.h (see Appendix A). You may, if you wish, simply include the file nr.h at the beginning of your own application program and use the recipes with abandon. That is what we have done in the example routines. It is somewhat more efficient, however, to cull from the file only those declarations which your program requires, and to declare them individually. You will find that the Numerical Recipes functions, in which we sought to make the interdependence of functions very clear, are written in this fashion.

The routine nrerror(), incidentally, forces a halt to the program with the function exit(). This function is technically not standard, but it appears in every C library we have checked. If you find yourself without this function, you may of course create it. Anything that forces the program to stop, gracefully or otherwise, will do.

Numerical Recipes
Example Book (C)

Chapter 1: Preliminaries

The routines in Chapter 1 of Numerical Recipes are introductory and less general in purpose than those in the remainder of the book. This chapter's routines serve primarily to expose the book's notational conventions, illustrate control structures, and perhaps to amuse. You may even find them useful, but we hope that you will not use badluk *for serious purposes.*

⋆   ⋆   ⋆   ⋆

Procedure flmoon calculates the phases of the moon, or more exactly, the Julian day and fraction thereof on which a given phase will occur or has occurred. The program xflmoon.c asks the present date and compiles a list of upcoming phases. We have compared the predictions to lunar tables, with happy results. Shown are the results of a test run, which you may replicate as a check. In this program, notice that we have set ZON (the time zone) to −5.0 to signify the five hour separation of the Eastern Standard time zone from Greenwich, England. Our convention requires you to use negative values of ZON if you are west of Greenwich, as we are. The Julian day results are converted to calendar dates through the use of caldat, which appears later in the chapter. The fractional Julian day and time zone combine to form a correction that can possibly change the calendar date by one day.

	Date		Time(EST)	Phase
1	9	1982	3 PM	full moon
1	16	1982	7 PM	last quarter
1	24	1982	11 PM	new moon
2	1	1982	10 AM	first quarter
2	8	1982	2 AM	full moon
2	15	1982	3 PM	last quarter
2	23	1982	4 PM	new moon
3	2	1982	6 PM	first quarter
3	9	1982	3 PM	full moon
3	17	1982	12 AM	last quarter
3	25	1982	5 AM	new moon
4	1	1982	0 AM	first quarter
4	8	1982	5 AM	full moon
4	16	1982	7 AM	last quarter
4	23	1982	4 PM	new moon
4	30	1982	7 AM	first quarter
5	7	1982	8 PM	full moon
5	16	1982	0 AM	last quarter
5	23	1982	0 AM	new moon
5	29	1982	3 PM	first quarter

```
/* Driver for routine FLMOON */

#include <stdio.h>
#include "nr.h"

#define ZON -5.0

main()
{
    int i,i1,i2,i3,id,im,iy,n,nph;
    float timzon = ZON/24.0,frac,secs;
    long j1,j2;
    static char *phase[]={"new moon","first quarter",
        "full moon","last quarter"};

    printf("Date of the next few phases of the moon\n");
    printf("Enter today\'s date (e.g. 1 31 1982)  :  \n");
    scanf("%d %d %d",&im,&id,&iy);
    /* Approximate number of full moons since january 1900 */
    n=12.37*(iy-1900+((im-0.5)/12.0));
    nph=2;
    j1=julday(im,id,iy);
    flmoon(n,nph,&j2,&frac);
    n += (float) (j1-j2)/28.0;
    printf("\n%10s %19s %9s\n","date","time(EST)","phase");
    for (i=1;i<=20;i++) {
        flmoon(n,nph,&j2,&frac);
        frac=24.0*(frac+timzon);
        if (frac < 0.0) {
            --j2;
            frac += 24.0;
        }
        if (frac > 12.0) {
            ++j2;
            frac -= 12.0;
        } else
            frac += 12.0;
        i1=(int) frac;
        secs=3600.0*(frac-i1);
        i2=(int) secs/60.0;
        i3=(int) (secs-60*i2+0.5);
        caldat(j2,&im,&id,&iy);
        printf("%5d %3d %5d %7d:%2d:%2d        %s\n",
            im,id,iy,i1,i2,i3,phase[nph]);
        if (nph == 3) {
            nph=0;
            ++n;
        } else
            ++nph;
    }
}
```

Program `julday`, our exemplar of the IF control structure, converts calendar dates to Julian dates. Not many people know the Julian date of their birthday or any other convenient reference point, for that matter. To remedy this, we offer a list of checkpoints, which appears at the end of this chapter as the file `dates1.dat`. The program `xjulday.c` lists the Julian date for each historic event for comparison. Then it allows you to make your own choices for entertainment.

```
/* Driver for JULDAY */

#include <stdio.h>
#include "nr.h"
#include "nrutil.h"

#define MAXSTR 80

main()
{
    int i,id,im,iy,n;
    char txt[MAXSTR];
    static char *name[]={"","january","february","march",
        "april","may","june","july","august","september",
        "october","november","december"};
    FILE *fp;

    if ((fp = fopen("dates1.dat","r")) == NULL)
        nrerror("Data file DATES1.DAT not found\n");
    fgets(txt,MAXSTR,fp);
    fscanf(fp,"%d %*s ",&n);
    printf("\n%5s %8s %6s %12s %9s\n","month","day","year",
        "julian day","event");
    for (i=1;i<=n;i++) {
        fscanf(fp,"%d %d %d ",&im,&id,&iy);
        fgets(txt,MAXSTR,fp);
        printf("%-10s %3d %6d %10ld %5s %s",name[im],id,iy,
            julday(im,id,iy)," ",txt);
    }
    fclose(fp);
    printf("\nYour choices: (negative to end)\n");
    printf("month day year (e.g. 1 13 1905)\n");
    for (i=1;i<=20;i++) {
        printf("\nmm dd yyyy ?\n");
        scanf("%d %d %d",&im,&id,&iy);
        if (im < 0) return;
        printf("julian day: %ld \n",julday(im,id,iy));
    }
}
```

The next program in *Numerical Recipes* is badluk, an infamous code that combines the best and worst instincts of man. We include no demonstration program for badluk, not because we fear it, but because it is self-contained, with sample results appearing in the text.

Chapter 1 closes with routine caldat, which illustrates no new points, but complements julday by doing conversions from Julian day number to the month, day, and year on which the given Julian day began. This offers an opportunity, grasped by the demonstration program xcaldat.c, to push dates through both julday and caldat in succession, to see if they survive intact. This, of course, tests only your authors' ability to make mistakes backwards as well as forwards, but we hope you will share our optimism that correct results here speak well for both routines. (We have checked them a bit more carefully in other ways.)

```
/* Driver for routine CALDAT */

#include <stdio.h>
#include "nr.h"
#include "nrutil.h"

#define MAXSTR 80

main()
{
    int i,id,idd,im,imm,iy,iyy,n;
    long j;
    char dummy[MAXSTR];
    static char *name[]={"","january","february","march",
        "april","may","june","july","august",
        "september","october","november","december"};
    FILE *fp;

    /* Check whether CALDAT properly undoes the operation of JULDAY */
    if ((fp = fopen("dates1.dat","r")) == NULL)
        nrerror("Data file DATES1.DAT not found\n");
    fgets(dummy,MAXSTR,fp);
    fscanf(fp,"%d %*s ",&n);
    printf("\n %14s %43s\n","original date:","reconstructed date");
    printf("%8s %5s %6s %15s %12s %5s %6s\n","month","day","year",
        "julian day","month","day","year");
    for (i=1;i<=n;i++) {
        fscanf(fp,"%d %d %d ",&im,&id,&iy);
        fgets(dummy,MAXSTR,fp);
        j=julday(im,id,iy);
        caldat(j,&imm,&idd,&iyy);
        printf("%10s %3d %6d %131d %16s %3d %6d\n",name[im],
            id,iy,j,name[imm],idd,iyy);
    }
    fclose(fp);
}
```

File dates1.dat:

```
List of dates for testing routines in Chapter 1
16 entries
12 31   -1 End of millennium
01 01    1 One day later
10 14 1582 Day before Gregorian calendar
10 15 1582 Gregorian calendar adopted
01 17 1706 Benjamin Franklin born
04 14 1865 Abraham Lincoln shot
04 18 1906 San Francisco earthquake
05 07 1915 Sinking of the Lusitania
07 20 1923 Pancho Villa assassinated
05 23 1934 Bonnie and Clyde eliminated
07 22 1934 John Dillinger shot
04 03 1936 Bruno Hauptman electrocuted
05 06 1937 Hindenburg disaster
07 26 1956 Sinking of the Andrea Doria
06 05 1976 Teton dam collapse
05 23 1968 Julian Day 2440000
```

Chapter 2: Linear Algebraic Equations

Numerical Recipes Chapter 2 begins the "true grit" of numerical analysis by considering the solution of linear algebraic equations. This is done first by Gauss-Jordan elimination (gaussj), and then by LU decomposition with forward and back substitution (ludcmp and lubksb). For singular or nearly singular matrices the best choice is singular value decomposition with back substitution (svdcmp and svbksb). Several linear systems of special form, represented by tridiagonal, Vandermonde, and Toeplitz matrices, may be treated with procedures tridag, vander, *and* toeplz *respectively. Linear systems with relatively few non-zero coefficients, so-called "sparse" matrices, are handled by routine* sparse.

★   ★   ★   ★

gaussj performs Gauss-Jordan elimination with full pivoting to find the solution of a set of linear equations for a collection of right-hand side vectors. The demonstration routine xgaussj.c checks its operation with reference to a group of test input matrices printed at the end of this chapter as file matrx1.dat. Each matrix is subjected to inversion by gaussj, and then multiplication by its own inverse to see that a unit matrix is produced. Then the solution vectors are each checked through multiplication by the original matrix and comparison with the right-hand side vectors that produced them.

```
/* Driver program for subroutine GAUSSJ */

#include <stdio.h>
#include "nr.h"
#include "nrutil.h"

#define NP 20
#define MP 20
#define MAXSTR 80

main()
{
    int j,k,l,m,n;
    float **a,**ai,**u,**b,**x,**t;
    char dummy[MAXSTR];
    FILE *fp;

    a=matrix(1,NP,1,NP);
    ai=matrix(1,NP,1,NP);
    u=matrix(1,NP,1,NP);
    b=matrix(1,NP,1,MP);
```

```
x=matrix(1,NP,1,MP);
t=matrix(1,NP,1,MP);
if ((fp = fopen("matrx1.dat","r")) == NULL)
    nrerror("Data file MATRX1.DAT not found\n");
while (!feof(fp)) {
    fgets(dummy,MAXSTR,fp);
    fgets(dummy,MAXSTR,fp);
    fscanf(fp,"%d %d ",&n,&m);
    fgets(dummy,MAXSTR,fp);
    for (k=1;k<=n;k++)
        for (l=1;l<=n;l++) fscanf(fp,"%f ",&a[k][l]);
    fgets(dummy,MAXSTR,fp);
    for (l=1;l<=m;l++)
        for (k=1;k<=n;k++) fscanf(fp,"%f ",&b[k][l]);
    /* save matrices for later testing of results */
    for (l=1;l<=n;l++) {
        for (k=1;k<=n;k++) ai[k][l]=a[k][l];
        for (k=1;k<=m;k++) x[l][k]=b[l][k];
    }
    /* invert matrix */
    gaussj(ai,n,x,m);
    printf("\nInverse of matrix a : \n");
    for (k=1;k<=n;k++) {
        for (l=1;l<=n;l++) printf("%12.6f",ai[k][l]);
        printf("\n");
    }
    /* check inverse */
    printf("\na times a-inverse:\n");
    for (k=1;k<=n;k++) {
        for (l=1;l<=n;l++) {
            u[k][l]=0.0;
            for (j=1;j<=n;j++)
                u[k][l] += (a[k][j]*ai[j][l]);
        }
        for (l=1;l<=n;l++) printf("%12.6f",u[k][l]);
        printf("\n");
    }
    /* check vector solutions */
    printf("\nCheck the following for equality:\n");
    printf("%21s %14s\n","original","matrix*sol'n");
    for (l=1;l<=m;l++) {
        printf("vector %2d: \n",l);
        for (k=1;k<=n;k++) {
            t[k][l]=0.0;
            for (j=1;j<=n;j++)
                t[k][l] += (a[k][j]*x[j][l]);
            printf("%8s %12.6f %12.6f\n"," ",
                b[k][l],t[k][l]);
        }
    }
    printf("***********************************\n");
    printf("press RETURN for next problem:\n");
    getchar();
}
fclose(fp);
free_matrix(t,1,NP,1,NP);
free_matrix(x,1,NP,1,NP);
```

```
        free_matrix(b,1,NP,1,NP);
        free_matrix(u,1,NP,1,NP);
        free_matrix(ai,1,NP,1,NP);
        free_matrix(a,1,NP,1,NP);
}
```

The demonstration program for routine `ludcmp` relies on the same package of test matrices, but just performs an LU decomposition of each. The performance is checked by multiplying the lower and upper matrices of the decomposition and comparing with the original matrix. The array `indx` keeps track of the scrambling done by `ludcmp` to effect partial pivoting. We had to do the unscrambling here, but you will normally not be called upon to do so, since `ludcmp` is used with the descrambler-containing routine `lubksb`.

```
/* Driver for routine LUDCMP */

#include <stdio.h>
#include "nr.h"
#include "nrutil.h"

#define NP 20
#define MAXSTR 80

main()
{
    int j,k,l,m,n,dum,*indx,*jndx;
    float d,**a,**xl,**xu,**x;
    char dummy[MAXSTR];
    FILE *fp;

    indx=ivector(1,NP);
    jndx=ivector(1,NP);
    a=matrix(1,NP,1,NP);
    xl=matrix(1,NP,1,NP);
    xu=matrix(1,NP,1,NP);
    x=matrix(1,NP,1,NP);
    if ((fp = fopen("matrx1.dat","r")) == NULL)
        nrerror("Data file MATRX1.DAT not found\n");
    while (!feof(fp)) {
        fgets(dummy,MAXSTR,fp);
        fgets(dummy,MAXSTR,fp);
        fscanf(fp,"%d %d ",&n,&m);
        fgets(dummy,MAXSTR,fp);
        for (k=1;k<=n;k++)
            for (l=1;l<=n;l++) fscanf(fp,"%f ",&a[k][l]);
        fgets(dummy,MAXSTR,fp);
        for (l=1;l<=m;l++)
            for (k=1;k<=n;k++) fscanf(fp,"%f ",&x[k][l]);
        /* Print out a-matrix for comparison with product of
           lower and upper decomposition matrices */
        printf("original matrix:\n");
        for (k=1;k<=n;k++) {
            for (l=1;l<=n;l++) printf("%12.6f",a[k][l]);
            printf("\n");
        }
        /* Perform the decomposition */
        ludcmp(a,n,indx,&d);
```

```
            /* Compose separately the lower and upper matrices */
            for (k=1;k<=n;k++) {
                for (l=1;l<=n;l++) {
                    if (l > k) {
                        xu[k][l]=a[k][l];
                        xl[k][l]=0.0;
                    } else if (l < k) {
                        xu[k][l]=0.0;
                        xl[k][l]=a[k][l];
                    } else {
                        xu[k][l]=a[k][l];
                        xl[k][l]=1.0;
                    }
                }
            }
            /* Compute product of lower and upper matrices for
               comparison with original matrix */
            for (k=1;k<=n;k++) {
                jndx[k]=k;
                for (l=1;l<=n;l++) {
                    x[k][l]=0.0;
                    for (j=1;j<=n;j++)
                        x[k][l] += (xl[k][j]*xu[j][l]);
                }
            }
            printf("\n%s%s\n","product of lower and upper ",
                "matrices (rows unscrambled):");
            for (k=1;k<=n;k++) {
                dum=jndx[indx[k]];
                jndx[indx[k]]=jndx[k];
                jndx[k]=dum;
            }
            for (k=1;k<=n;k++)
                for (j=1;j<=n;j++)
                    if (jndx[j] == k) {
                        for (l=1;l<=n;l++)
                            printf("%12.6f",x[j][l]);
                        printf("\n");
                    }
            printf("\nlower matrix of the decomposition:\n");
            for (k=1;k<=n;k++) {
                for (l=1;l<=n;l++) printf("%12.6f",xl[k][l]);
                printf("\n");
            }
            printf("\nupper matrix of the decomposition:\n");
            for (k=1;k<=n;k++) {
                for (l=1;l<=n;l++) printf("%12.6f",xu[k][l]);
                printf("\n");
            }
            printf("\n*********************************\n");
            printf("press return for next problem:\n");
            getchar();
        }
    fclose(fp);
    free_matrix(x,1,NP,1,NP);
    free_matrix(xu,1,NP,1,NP);
    free_matrix(xl,1,NP,1,NP);
```

```
        free_matrix(a,1,NP,1,NP);
        free_ivector(jndx,1,NP);
        free_ivector(indx,1,NP);
}
```

Our example driver for `lubksb` makes calls to both `ludcmp` and `lubksb` in order to solve the linear equation problems posed in file `matrx1.dat` (see discussion of `gaussj`). The original matrix of coefficients is applied to the solution vectors to check that the result matches the right-hand side vectors posed for each problem. We apologize for using routine `ludcmp` in a test of `lubksb`, but `ludcmp` has been tested independently, and anyway, `lubksb` is nothing without this partner program, so a test of the combination is more to the point.

```c
/* Driver for routine LUBKSB */

#include <stdio.h>
#include "nr.h"
#include "nrutil.h"

#define NP 20
#define MAXSTR 80

main()
{
    int j,k,l,m,n,*indx;
    float p,*x,**a,**b,**c;
    char dummy[MAXSTR];
    FILE *fp;

    indx=ivector(1,NP);
    x=vector(1,NP);
    a=matrix(1,NP,1,NP);
    b=matrix(1,NP,1,NP);
    c=matrix(1,NP,1,NP);
    if ((fp = fopen("matrx1.dat","r")) == NULL)
        nrerror("Data file MATRX1.DAT not found\n");
    while (!feof(fp)) {
        fgets(dummy,MAXSTR,fp);
        fgets(dummy,MAXSTR,fp);
        fscanf(fp,"%d %d ",&n,&m);
        fgets(dummy,MAXSTR,fp);
        for (k=1;k<=n;k++)
            for (l=1;l<=n;l++) fscanf(fp,"%f ",&a[k][l]);
        fgets(dummy,MAXSTR,fp);
        for (l=1;l<=m;l++)
            for (k=1;k<=n;k++) fscanf(fp,"%f ",&b[k][l]);
        /* Save matrix a for later testing */
        for (l=1;l<=n;l++)
            for (k=1;k<=n;k++) c[k][l]=a[k][l];
        /* Do lu decomposition */
        ludcmp(c,n,indx,&p);
        /* Solve equations for each right-hand vector */
        for (k=1;k<=m;k++) {
            for (l=1;l<=n;l++) x[l]=b[l][k];
            lubksb(c,n,indx,x);
            /* Test results with original matrix */
            printf("right-hand side vector:\n");
```

```
        for (l=1;l<=n;l++)
            printf("%12.6f",b[l][k]);
        printf("\n%s%s\n","result of matrix applied",
            " to sol'n vector");
        for (l=1;l<=n;l++) {
            b[l][k]=0.0;
            for (j=1;j<=n;j++)
                b[l][k] += (a[l][j]*x[j]);
        }
        for (l=1;l<=n;l++)
            printf("%12.6f",b[l][k]);
        printf("\n*******************************\n");
    }
    printf("press RETURN for next problem:\n");
    getchar();
}
fclose(fp);
free_matrix(c,1,NP,1,NP);
free_matrix(b,1,NP,1,NP);
free_matrix(a,1,NP,1,NP);
free_vector(x,1,NP);
free_ivector(indx,1,NP);
}
```

Procedure `tridag` solves linear equations with coefficients that form a tridiagonal matrix. We provide at the end of this chapter a second file of matrices `matrix2.dat` for the demonstration driver. In all other respects, the demonstration program `xtridag.c` operates in the same fashion as `xlubksb.c`.

```
/* Driver for routine TRIDAG */

#include <stdio.h>
#include "nr.h"
#include "nrutil.h"

#define NP 20
#define MAXSTR 80

main()
{
    int k,n;
    float *diag,*superd,*subd,*rhs,*u;
    char dummy[MAXSTR];
    FILE *fp;

    diag=vector(1,NP);
    superd=vector(1,NP);
    subd=vector(1,NP);
    rhs=vector(1,NP);
    u=vector(1,NP);
    if ((fp = fopen("matrx2.dat","r")) == NULL)
        nrerror("Data file MATRX2.DAT not found\n");
    while (!feof(fp)) {
        fgets(dummy,MAXSTR,fp);
        fgets(dummy,MAXSTR,fp);
        fscanf(fp,"%d ",&n);
        fgets(dummy,MAXSTR,fp);
```

```
    for (k=1;k<=n;k++) fscanf(fp,"%f ",&diag[k]);
    fgets(dummy,MAXSTR,fp);
    for (k=1;k<n;k++) fscanf(fp,"%f ",&superd[k]);
    fgets(dummy,MAXSTR,fp);
    for (k=2;k<=n;k++) fscanf(fp,"%f ",&subd[k]);
    fgets(dummy,MAXSTR,fp);
    for (k=1;k<=n;k++) fscanf(fp,"%f ",&rhs[k]);
    /* carry out solution */
    tridag(subd,diag,superd,rhs,u,n);
    printf("\nThe solution vector is:\n");
    for (k=1;k<=n;k++) printf("%12.6f",u[k]);
    printf("\n");
    /* test solution */
    printf("\n(matrix)*(sol'n vector) should be:\n");
    for (k=1;k<=n;k++) printf("%12.6f",rhs[k]);
    printf("\n");
    printf("actual result is:\n");
    for (k=1;k<=n;k++) {
        if (k == 1)
            rhs[k]=diag[1]*u[1]+superd[1]*u[2];
        else if (k == n)
            rhs[k]=subd[n]*u[n-1]+diag[n]*u[n];
        else
            rhs[k]=subd[k]*u[k-1]+diag[k]*u[k]
                +superd[k]*u[k+1];
    }
    for (k=1;k<=n;k++) printf("%12.6f",rhs[k]);
    printf("\n");
    printf("**********************************\n");
    printf("press return for next problem:\n");
    getchar();
    }
    fclose(fp);
    free_vector(u,1,NP);
    free_vector(rhs,1,NP);
    free_vector(subd,1,NP);
    free_vector(superd,1,NP);
    free_vector(diag,1,NP);
}
```

mprove is a short routine for improving the solution vector for a set of linear equations, providing that an LU decomposition has been performed on the matrix of coefficients. Our test of this function is to use ludcmp and lubksb to solve a set of equations specified by the initializations at the beginning of the program. The solution vector is then corrupted by the addition of random values to each component. mprove works on the corrupted vector to recover the original. Note the use of the utility routine convert_matrix() to change a conventionally defined matrix to the format used in *Numerical Recipes.*

```
/* Driver for routine MPROVE */

#include <stdio.h>
#include "nr.h"
#include "nrutil.h"

#define N 5
#define NP N
```

```
main()
{
    int i,idum,j,*indx;
    float d,*x,**a,**aa;
    static float ainit[NP][NP]=
        {1.0,2.0,3.0,4.0,5.0,
         2.0,3.0,4.0,5.0,1.0,
         1.0,1.0,1.0,1.0,1.0,
         4.0,5.0,1.0,2.0,3.0,
         5.0,1.0,2.0,3.0,4.0};
    static float b[N+1]={0.0,1.0,1.0,1.0,1.0,1.0};

    indx=ivector(1,N);
    x=vector(1,N);
    a=convert_matrix(&ainit[0][0],1,N,1,N);
    aa=matrix(1,N,1,N);
    for (i=1;i<=N;i++) {
        x[i]=b[i];
        for (j=1;j<=N;j++)
            aa[i][j]=a[i][j];
    }
    ludcmp(aa,N,indx,&d);
    lubksb(aa,N,indx,x);
    printf("\nSolution vector for the equations:\n");
    for (i=1;i<=N;i++) printf("%12.6f",x[i]);
    printf("\n");
    /* now phoney up x and let MPROVE fix it */
    idum = -13;
    for (i=1;i<=N;i++) x[i] *= (1.0+0.2*ran3(&idum));
    printf("\nSolution vector with noise added:\n");
    for (i=1;i<=N;i++) printf("%12.6f",x[i]);
    printf("\n");
    mprove(a,aa,N,indx,b,x);
    printf("\nSolution vector recovered by MPROVE:\n");
    for (i=1;i<=N;i++) printf("%12.6f",x[i]);
    printf("\n");
    free_matrix(aa,1,N,1,N);
    free_convert_matrix(a,1,N,1,N);
    free_vector(x,1,N);
    free_ivector(indx,1,N);
}
```

Vandermonde matrices of dimension $N \times N$ have elements that are entirely integer powers of N arbitrary numbers $x_1 \ldots x_N$. (See *Numerical Recipes* for details). In the demonstration program xvander.c we provide five such numbers to specify a 5×5 matrix, and five elements of a right-hand side vector Q. Routine vander is used to find the solution vector W. This vector is tested by applying the matrix to W and comparing the result to Q.

```
/* Driver for routine VANDER */

#include <stdio.h>
#include "nr.h"
#include "nrutil.h"

#define N 5
```

```
main()
{
    int i,j;
    float sum,*w,*term;
    static float x[]={0.0,1.0,1.5,2.0,2.5,3.0};
    static float q[]={0.0,1.0,1.5,2.0,2.5,3.0};

    w=vector(1,N);
    term=vector(1,N);
    vander(x,w,q,N);
    printf("\nSolution vector:\n");
    for (i=1;i<=N;i++)
        printf("%7s%1d%2s %12f \n","w[",i,"]=",w[i]);
    printf("\nTest of solution vector:\n");
    printf("%14s %11s\n","mtrx*sol'n","original");
    sum=0.0;
    for (i=1;i<=N;i++) {
        term[i]=w[i];
        sum += w[i];
    }
    printf("%12.4f %12.4f\n",sum,q[1]):
    for (i=2;i<=N;i++) {
        sum=0.0;
        for (j=1;j<=N;j++) {
            term[j] *= x[j];
            sum += term[j];
        }
        printf("%12.4f %12.4f\n",sum,q[i]);
    }
    free_vector(term,1,N);
    free_vector(w,1,N);
}
```

A very similar test is applied to `toeplz`, which operates on Toeplitz matrices. The $N \times N$ Toeplitz matrix is specified by $2N-1$ numbers r_i, in this case taken to be simply a linear progression of values. A right-hand side y_i is chosen likewise. `toeplz` finds the solution vector x_i, and checks it in the usual fashion.

```
/* Driver for routine TOEPLZ */

#include <stdio.h>
#include "nr.h"

#define N 5
#define TWON (2*N)

main()
{
    int i,j;
    float sum,r[TWON+1],x[N+1],y[N+1];

    for (i=1;i<=N;i++) y[i]=0.1*i;
    for (i=1;i<TWON;i++) r[i]=0.1*i;
    toeplz(r,x,y,N);
    printf("Solution vector:\n");
    for (i=1;i<=N;i++)
```

```
        printf("%7s%1d%s %13f\n","x[",i,"] =",x[i]);
    printf("\nTest of solution:\n");
    printf("%13s %12s\n","mtrx*soln","original");
    for (i=1;i<=N;i++) {
        sum=0.0;
        for (j=1;j<=N;j++)
            sum += (r[N+i-j]*x[j]);
        printf("%12.4f %12.4f\n",sum,y[i]);
    }
}
```

The pair svdcmp, svbksb are tested in the same manner as ludcmp, lubksb. That is, svdcmp is checked independently to see that it yields proper decomposition of matrices. Then the pair of programs is tested as a unit to see that they provide correct solutions to some linear sets. (Note: Because of the order of programs in *Numerical Recipes*, the test of the pair in this case comes first). The matrices and solution vectors are given in the Appendix as file matrx3.dat.

Driver xsvbksb.c brings in matrices a and right-hand side vectors b from ma-trx3.dat. Matrix a, itself, is saved for later use. It is copied into matrix u for processing by svdcmp. The results of the processing are the three arrays u, w, v which form the singular value decomposition of a. The right-hand side vectors are fed one at a time to vector c, and the resulting solution vectors x are checked for accuracy through application of the saved matrix a.

```
/* Driver for routine SVBKSB */

#include <stdio.h>
#include "nr.h"
#include "nrutil.h"

#define NP 20
#define MP 20
#define MAXSTR 80

main()
{
    int j,k,l,m,n;
    float wmax,wmin,*w,*x,*c;
    float **a,**b,**u,**v;
    char dummy[MAXSTR];
    FILE *fp;

    w=vector(1,NP);
    x=vector(1,NP);
    c=vector(1,MP);
    a=matrix(1,MP,1,NP);
    b=matrix(1,MP,1,NP);
    u=matrix(1,MP,1,NP);
    v=matrix(1,NP,1,NP);
    if ((fp = fopen("matrx1.dat","r")) == NULL)
        nrerror("Data file MATRX1.DAT not found\n");
    while (!feof(fp)) {
        fgets(dummy,MAXSTR,fp);
        fgets(dummy,MAXSTR,fp);
        fscanf(fp,"%d %d ",&n,&m);
```

```
    fgets(dummy,MAXSTR,fp);
    for (k=1;k<=n;k++)
        for (l=1;l<=n;l++) fscanf(fp,"%f ",&a[k][l]);
    fgets(dummy,MAXSTR,fp);
    for (l=1;l<=m;l++)
        for (k=1;k<=n;k++) fscanf(fp,"%f ",&b[k][l]);
    /* copy a into u */
    for (k=1;k<=n;k++)
        for (l=1;l<=n;l++) u[k][l]=a[k][l];
    /* decompose matrix a */
    svdcmp(u,n,n,w,v);
    /* find maximum singular value */
    wmax=0.0;
    for (k=1;k<=n;k++)
        if (w[k] > wmax) wmax=w[k];
    /* define "small" */
    wmin=wmax*(1.0e-6);
    /* zero the "small" singular values */
    for (k=1;k<=n;k++)
        if (w[k] < wmin) w[k]=0.0;
    /* backsubstitute for each right-hand side vector */
    for (l=1;l<=m;l++) {
        printf("\nVector number %2d\n",l);
        for (k=1;k<=n;k++) c[k]=b[k][l];
        svbksb(u,w,v,n,n,c,x);
        printf(" solution vector is:\n");
        for (k=1;k<=n;k++) printf("%12.6f",x[k]);
        printf("\n original right-hand side vector:\n");
        for (k=1;k<=n;k++) printf("%12.6f",c[k]);
        printf("\n (matrix)*(sol'n vector):\n");
        for (k=1;k<=n;k++) {
            c[k]=0.0;
            for (j=1;j<=n;j++)
                c[k] += a[k][j]*x[j];
        }
        for (k=1;k<=n;k++) printf("%12.6f",c[k]);
        printf("\n");
    }
    printf ("**********************************\n");
    printf("press RETURN for next problem\n");
    getchar();
}
fclose(fp);
free_matrix(v,1,NP,1,NP);
free_matrix(u,1,MP,1,NP);
free_matrix(b,1,MP,1,NP);
free_matrix(a,1,MP,1,NP);
free_vector(c,1,MP);
free_vector(x,1,NP);
free_vector(w,1,NP);
}
```

Companion driver xsvdcmp.c takes the same matrices from matrx3.dat and passes copies u to svdcmp for singular value decomposition into u, w, and v. Then u, w, and the transpose of v are multiplied together. The result is compared to a saved copy of a.

```
/* Driver for routine SVDCMP */

#include <stdio.h>
#include "nr.h"
#include "nrutil.h"

#define NP 20
#define MP 20
#define MAXSTR 80

main()
{
    int j,k,l,m,n;
    float *w,**a,**u,**v;
    char dummy[MAXSTR];
    FILE *fp;

    w=vector(1,NP);
    a=matrix(1,MP,1,NP);
    u=matrix(1,MP,1,NP);
    v=matrix(1,NP,1,NP);
    /* read input matrices */
    if ((fp = fopen("matrx3.dat","r")) == NULL)
        nrerror("Data file MATRX3.DAT not found\n");
    while (!feof(fp)) {
        fgets(dummy,MAXSTR,fp);
        fgets(dummy,MAXSTR,fp);
        fscanf(fp,"%d %d ",&m,&n);
        fgets(dummy,MAXSTR,fp);
        /* copy original matrix into u */
        for (k=1;k<=m;k++)
            for (l=1;l<=n;l++) {
                fscanf(fp,"%f ",&a[k][l]);
                u[k][l]=a[k][l];
            }
        if (n > m) {
            for (k=m+1;k<=n;k++) {
                for (l=1;l<=n;l++) {
                    a[k][l]=0.0;
                    u[k][l]=0.0;
                }
            }
            m=n;
        }
        /* perform decomposition */
        svdcmp(u,m,n,w,v);
        /* write results */
        printf("Decomposition matrices:\n");
        printf("Matrix u\n");
        for (k=1;k<=m;k++) {
            for (l=1;l<=n;l++)
                printf("%12.6f",u[k][l]);
            printf("\n");
        }
        printf("Diagonal of matrix w\n");
        for (k=1;k<=n;k++)
            printf("%12.6f",w[k]);
```

```
            printf("\nMatrix v-transpose\n");
            for (k=1;k<=n;k++) {
                for (l=1;l<=n;l++)
                    printf("%12.6f",v[l][k]);
                printf("\n");
            }
            printf("\nCheck product against original matrix:\n");
            printf("Original matrix:\n");
            for (k=1;k<=m;k++) {
                for (l=1;l<=n;l++)
                    printf("%12.6f",a[k][l]);
                printf("\n");
            }
            printf("Product u*w*(v-transpose):\n");
            for (k=1;k<=m;k++) {
                for (l=1;l<=n;l++) {
                    a[k][l]=0.0;
                    for (j=1;j<=n;j++)
                        a[k][l] += u[k][j]*w[j]*v[l][j];
                }
                for (l=1;l<=n;l++) printf("%12.6f",a[k][l]);
                printf("\n");
            }
            printf("***********************************\n");
            printf("press RETURN for next problem\n");
            getchar();
    }
    fclose(fp);
    free_matrix(v,1,NP,1,NP);
    free_matrix(u,1,MP,1,NP);
    free_matrix(a,1,MP,1,NP);
    free_vector(w,1,NP);
}
```

Routine sparse solves linear systems $\mathbf{A} \cdot \mathbf{x} = \mathbf{b}$ with a sparse matrix $\mathbf{A}$. Rather than specifying the entire matrix $\mathbf{A}$ (most elements of which are zero), the program calls two procedures asub and atsub which are, for any input vector xin, supposed to return the result xout of applying $\mathbf{A}$ and its transpose to xin, respectively. In our sample program we define these two procedures to implement the 20×20 matrix

$$\begin{pmatrix} 1.0 & 2.0 & 0.0 & 0.0 & \ldots \\ -2.0 & 1.0 & 2.0 & 0.0 & \ldots \\ 0.0 & -2.0 & 1.0 & 2.0 & \ldots \\ 0.0 & 0.0 & -2.0 & 1.0 & \ldots \\ \vdots & \vdots & \vdots & \vdots & \ddots \end{pmatrix}$$

As a right-hand side vector $\mathbf{b}$ we have taken $(3.0, 1.0, 1.0, \ldots, -1.0)$, and the solution is given as $\mathbf{x}$. Notice that the components of $\mathbf{x}$ are all initialized to zero. You will set them to some initial guess of the solution to your own problem, but this guess will usually suffice. The solution in xsparse.c is given the usual checks.

```
/* Driver for routine SPARSE */

#include <stdio.h>
#include "nr.h"
```

```
#define N 20

void asub(xin,xout,n)
float xin[],xout[];
int n;
{
    int i;

    xout[1]=xin[1]+2.0*xin[2];
    xout[n] = -2.0*xin[n-1]+xin[n];
    for (i=2;i<n;i++)
        xout[i] = -2.0*xin[i-1]+xin[i]+2.0*xin[i+1];
}

void atsub(xin,xout,n)
float xin[],xout[];
int n;
{
    int i;

    xout[1]=xin[1]-2.0*xin[2];
    xout[n]=2.0*xin[n-1]+xin[n];
    for (i=2;i<n;i++)
        xout[i]=2.0*xin[i-1]+xin[i]-2.0*xin[i+1];
}

main()
{
    int i,ii;
    float rsq;
    float b[N+1],bcmp[N+1],x[N+1];

    for (i=1;i<=N;i++) {
        x[i]=0.0;
        b[i]=1.0;
    }
    b[1]=3.0;
    b[N] = -1.0;
    sparse(b,N,x,&rsq);
    printf("%s %15f\n","sum-squared residual:",rsq);
    printf("\nsolution vector:\n");
    for (ii=1;ii<=N/5;ii++)
    {
        for (i=5*(ii-1)+1;i<=5*ii;i++) printf("%12.6f",x[i]);
        printf("\n");
    }
    for (i=1;i<=(N % 5);i++)
        printf("%12.6f",x[5*(N/5)+i]);
    printf("\n");
    asub(x,bcmp,N);
    printf("\npress RETURN to continue...\n");
    getchar();
    printf("test of solution vector:\n");
    printf("%9s %12s\n","a*x","b");
    for (i=1;i<=N;i++)
        printf("%12.6f %12.6f\n",bcmp[i],b[i]);
}
```

Appendix

File matrx1.dat:

```
MATRICES FOR INPUT TO TEST ROUTINES
Size of matrix (NxN), Number of solutions:
3 2
Matrix A:
1.0 0.0 0.0
0.0 2.0 0.0
0.0 0.0 3.0
Solution vectors:
1.0 0.0 0.0
1.0 1.0 1.0
NEXT PROBLEM
Size of matrix (NxN), Number of solutions:
3 2
Matrix A:
1.0 2.0 3.0
2.0 2.0 3.0
3.0 3.0 3.0
Solution vectors:
1.0 1.0 1.0
1.0 2.0 3.0
NEXT PROBLEM:.
Size of matrix (NxN), Number of solutions:
5 2
Matrix A:
1.0 2.0 3.0 4.0 5.0
2.0 3.0 4.0 5.0 1.0
3.0 4.0 5.0 1.0 2.0
4.0 5.0 1.0 2.0 3.0
5.0 1.0 2.0 3.0 4.0
Solution vectors:
1.0 1.0 1.0 1.0 1.0
1.0 2.0 3.0 4.0 5.0
NEXT PROBLEM:
Size of matrix (NxN), Number of solutions:
5 2
Matrix A:
1.4 2.1 2.1 7.4 9.6
1.6 1.5 1.1 0.7 5.0
3.8 8.0 9.6 5.4 8.8
4.6 8.2 8.4 0.4 8.0
2.6 2.9 0.1 9.6 7.7
Solution vectors:
1.1 1.6 4.7 9.1 0.1
4.0 9.3 8.4 0.4 4.1
```

File matrx2.dat:

```
FILE OF TRIDIAGONAL MATRICES FOR PROGRAM 'TRIDAG'
Dimension of matrix
3
Diagonal elements (N)
1.0 2.0 3.0
Super-diagonal elements (N-1)
2.0 3.0
```

```
Sub-diagonal elements (N-1)
2.0 3.0
Right-hand side vector (N)
1.0 2.0 3.0
NEXT PROBLEM:
Dimension of matrix
5
Diagonal elements (N)
1.0 1.0 1.0 1.0 1.0
Super-diagonal elements (N-1)
1.0 2.0 3.0 4.0
Sub-diagonal elements (N-1)
2.0 3.0 4.0 5.0
Right-hand side vector (N)
1.0 2.0 3.0 4.0 5.0
NEXT PROBLEM:
Dimension of matrix
5
Diagonal elements (N)
1.0 2.0 3.0 4.0 5.0
Super-diagonal elements (N-1)
2.0 3.0 4.0 5.0
Sub-diagonal elements (N-1)
2.0 3.0 4.0 5.0
Right-hand side vector (N)
1.0 1.0 1.0 1.0 1.0
NEXT PROBLEM:
Dimension of matrix
6
Diagonal elements (N)
9.7 9.5 5.2 3.5 5.1 6.0
Super-diagonal elements (N-1)
6.0 1.2 0.7 3.0 1.5
Sub-diagonal elements (N-1)
2.1 9.4 3.3 7.5 8.8
Right-hand side vector (N)
2.0 7.5 0.6 7.4 9.8 8.8
```

File matrx3.dat:

```
FILE OF MATRICES FOR SVDCMP:
Number of Rows, Columns
5 3
Matrix
1.0 2.0 3.0
2.0 3.0 4.0
3.0 4.0 5.0
4.0 5.0 6.0
5.0 6.0 7.0
NEXT PROBLEM:
Number of Rows, Columns
5 5
Matrix
1.0 2.0 3.0 4.0 5.0
2.0 2.0 3.0 4.0 5.0
3.0 3.0 3.0 4.0 5.0
4.0 4.0 4.0 4.0 5.0
5.0 5.0 5.0 5.0 5.0
```

```
NEXT PROBLEM:
Number of Rows, Columns
6 6
Matrix
3.0 5.3 5.6 3.5 6.8 5.7
0.4 8.2 6.7 1.9 2.2 5.3
7.8 8.3 7.7 3.3 1.9 4.8
5.5 8.8 3.0 1.0 5.1 6.4
5.1 5.1 3.6 5.8 5.7 4.9
3.5 2.7 5.7 8.2 9.6 2.9
```

Chapter 3: Interpolation and Extrapolation

Chapter 3 of *Numerical Recipes* deals with interpolation and extrapolation (the same routines are usable for both). Three fundamental interpolation methods are first discussed,

1. *Polynomial interpolation* (polint),

2. *Rational function interpolation* (ratint), and

3. *Cubic spline interpolation* (spline, splint).

To find the place in an ordered table at which to perform an interpolation, two routines are given, locate and hunt. Also, for cases in which the actual coefficients of a polynomial interpolation are desired, the routines polcoe and polcof are provided (along with important warnings circumscribing their usefulness).

For higher-dimensional interpolations, *Numerical Recipes* treats only problems on a regularly spaced grid. Routine polin2 does a two-dimensional polynomial interpolation that aims at accuracy rather than smoothness. When smooth interpolation is desired, the methods shown in bcucof and bcuint for bicubic interpolation are recommended. In the case of two-dimensional spline interpolations, the routines splie2 and splin2 are offered.

$\star$ $\quad$ $\star$ $\quad$ $\star$ $\quad$ $\star$

Program polint takes two arrays xa and ya of length N that express the known values of a function, and calculates the value, at a point x, of the unique polynomial of degree $N - 1$ passing through all the given values. For the purpose of illustration, in xpolint.c we have taken evenly spaced xa[i] and set ya[i] equal to simple functions (sines and exponentials) of these xa[i]. For the sine we use an interval of length π, and for the exponential an interval of length 1.0. You may choose the number N of reference points and observe the improvement of the results as N increases. The test points x are slightly shifted from the reference points so that you can compare the estimated error dy with the actual error. By removing the shift, you may check that the polynomial actually hits all reference points.

```
/* Driver for routine POLINT */

#include <stdio.h>
#include <math.h>
#include "nr.h"
#include "nrutil.h"
```

```
#define NP 10      /* maximum value of n */
#define PI 3.1415926

main()
{
    int  i,n,nfunc;
    float  dy,f,x,y,*xa,*ya;

    xa=vector(1,NP);
    ya=vector(1,NP);
    printf("generation of interpolation tables\n");
    printf(" ... sin(x)      0<x<PI\n");
    printf(" ... exp(x)      0<x<1 \n");
    printf("how many entries go in these tables? (note: n<10)\n");
    scanf("%d",&n);
    for (nfunc=1;nfunc<=2;nfunc++) {
        if (nfunc == 1)     {
            printf("\nsine function from 0 to PI\n");
            for (i=1;i<=n;i++) {
                xa[i]=i*PI/n;
                ya[i]=sin(xa[i]);
            }
        } else if (nfunc == 2) {
            printf("\nexponential function from 0 to 1\n");
            for (i=1;i<=n;i++) {
                xa[i]=i*1.0/n;
                ya[i]=exp(xa[i]);
            }
        } else {
            free_vector(ya,1,NP);
            free_vector(xa,1,NP);
            return;
        }
        printf("\n%9s %13s %16s %13s\n",
            "x","f(x)","interpolated","error");
        for (i=1;i<=10;i++) {
            if (nfunc == 1)     {
                x=(-0.05+i/10.0)*PI;
                f=sin(x);
            } else if (nfunc == 2) {
                x=(-0.05+i/10.0);
                f=exp(x);
            }
            polint(xa,ya,n,x,&y,&dy);
            printf("%12.6f %12.6f %12.6f %4s %11f\n",
                x,f,y," ",dy);
        }
        printf("\n**********************************\n");
        printf("press RETURN\n");
        getchar();
    }
}
```

.ratint is functionally similar to polint in that it also returns a value y for the function at point x, and an error estimate dy as well. In this case the values are determined from the unique diagonal rational function that passes through all the reference points. If you inspect the driver closely, you will find that two of the test points fall directly on top of reference points and should give exact results. The remainder do not. You can compare the estimated error dyy to the actual error $|yy - yexp|$ for these cases.

```
/* Driver for routine RATINT */

#include <stdio.h>
#include <math.h>
#include "nr.h"
#include "nrutil.h"

#define NPT 6
#define EPS 1.0
#define SQR(a) ((a)*(a))

float f(x,eps)
float x,eps;
{
    return x*exp(-x)/(SQR(x-1.0)+SQR(eps));
}

main()
{
    int i;
    float dyy,xx,yexp,yy,*x,*y;

    x=vector(1,NPT);
    y=vector(1,NPT);
    for (i=1;i<=NPT;i++) {
        x[i]=i*2.0/NPT;
        y[i]=f(x[i],EPS);
    }
    printf("\nDiagonal rational function interpolation\n");
    printf("\n%5s %13s %14s %12s\n","x","interp.","accuracy","actual");
    for (i=1;i<=10;i++) {
        xx=0.2*i;
        ratint(x,y,NPT,xx,&yy,&dyy);
        yexp=f(xx,EPS);
        printf("%6.2f %12.6f    %11f %13.6f\n",xx,yy,dyy,yexp);
    }
    free_vector(y,1,NPT);
    free_vector(x,1,NPT);
}
```

Procedure spline generates a cubic spline. Given an array of x_i and $f(x_i)$, and given values of the first derivative of function f at the two endpoints of the tabulated region, it returns the second derivative of f at each of the tabulation points. As an example we chose the function $\sin x$ and evaluated it at evenly spaced points x[i]. In this case the first derivatives at the end-points are ypl $= \cos x_1$ and ypn $= \cos x_N$. The output array of spline is y2[i] and this is listed along with $-\sin x_i$, the second derivative of $\sin x_i$, for comparison.

```
/* Driver for routine SPLINE */

#include <stdio.h>
#include <math.h>
#include "nr.h"
#include "nrutil.h"
```

```
#define N 20
#define PI 3.1415926

main()
{
    int i;
    float yp1,ypn,*x,*y,*y2;

    x=vector(1,N);
    y=vector(1,N);
    y2=vector(1,N);
    printf("\nsecond-derivatives for sin(x) from 0 to pi\n");
    /* Generate array for interpolation */
    for (i=1;i<=20;i++) {
        x[i]=i*PI/N;
        y[i]=sin(x[i]);
    }
    /* calculate 2nd derivative with spline */
    yp1=cos(x[1]);
    ypn=cos(x[N]);
    spline(x,y,N,yp1,ypn,y2);
    /* test result */
    printf("%23s %16s\n","spline","actual");
    printf("%11s %14s %16s\n","number","2nd deriv","2nd deriv");
    for (i=1;i<=N;i++)
        printf("%8d %16.6f %16.6f\n",i,y2[i],-sin(x[i]));
    free_vector(y2,1,N);
    free_vector(y,1,N);
    free_vector(x,1,N);
}
```

Actual cubic-spline interpolations, however, are carried out by `splint`. This routine uses the output array from one call to `spline` to service any subsequent number of spline interpolations with different x's. The demonstration program `xsplint.c` tests this capability on both $\sin x$ and $\exp x$. The two are treated in succession according to whether `nfunc` is one or two. In each case the function is tabulated at equally spaced points, and the derivatives are found at the first and last point. A call to `spline` then produces an array of second derivatives `y2` which is fed to `splint`. The interpolated values `y` are compared with actual function values `f` at a different set of equally spaced points.

```
/* Driver for routine SPLINT */

#include <stdio.h>
#include <math.h>
#include "nr.h"
#include "nrutil.h"

#define NP 10
#define PI 3.1415926

main()
{
    int i,nfunc;
    float f,x,y,yp1,ypn,*xa,*ya,*y2;
```

```
    xa=vector(1,NP);
    ya=vector(1,NP);
    y2=vector(1,NP);
    for (nfunc=1;nfunc<=2;nfunc++) {
        if (nfunc == 1) {
            printf("\nsine function from 0 to pi\n");
            for (i=1;i<=NP;i++) {
                xa[i]=i*PI/NP;
                ya[i]=sin(xa[i]);
            }
            yp1=cos(xa[1]);
            ypn=cos(xa[NP]);
        } else if (nfunc == 2) {
            printf("\nexponential function from 0 to 1\n");
            for (i=1;i<=NP;i++) {
                xa[i]=1.0*i/NP;
                ya[i]=exp(xa[i]);
            }
            yp1=exp(xa[1]);
            ypn=exp(xa[NP]);
        } else {
            free_vector(y2,1,NP);
            free_vector(ya,1,NP);
            free_vector(xa,1,NP);
            return;
        }
        /* Call spline to get second derivatives */
        spline(xa,ya,NP,yp1,ypn,y2);
        /* Call splint for interpolations */
        printf("\n%9s %13s %17s\n","x","f(x)","interpolation");
        for (i=1;i<=10;i++) {
            if (nfunc == 1) {
                x=(-0.05+i/10.0)*PI;
                f=sin(x);
            } else if (nfunc == 2) {
                x = -0.05+i/10.0;
                f=exp(x);
            }
            splint(xa,ya,y2,NP,x,&y);
            printf("%12.6f %12.6f %12.6f\n",x,f,y);
        }
        printf("\n***********************************\n");
        printf("Press RETURN\n");
        getchar();
    }
}
```

The next program, `locate`, may be used in conjunction with any interpolation method to bracket the x-position for which $f(x)$ is sought by two adjacent tabulated positions. That is, given a monotonic array of x_i, and given a value of x, it finds the two values x_i, x_{i+1} that surround x. In xlocate.c we chose the array x_i to be non-uniform, varying exponentially with i. Then we took a uniform series of x-values and sought their position in the array using `locate`. For each x, `locate` finds the value j for which x[j] is nearest below x. Then the driver shows j, and the two bracketing values xx[j] and xx[j+1]. If j is 0 or n, then x is not within the tabulated range.

The program thereby flags 'lower lim' if x is below x[1] or 'upper lim' if x is above x[n].

```
/* Driver for routine LOCATE */

#include <stdio.h>
#include <math.h>
#include "nr.h"
#include "nrutil.h"

#define N 100

main()
{
    int i,j;
    float x,*xx;

    xx=vector(1,N);
    /* create array to be searched */
    for (i=1;i<=N;i++)
        xx[i]=exp(i/20.0)-74.0;
    printf("\nresult of:  j=0 indicates x too small\n");
    printf("%11s j=100 indicates x too large"," ");
    printf("\n%10s %6s %11s %12s \n","locate ","j","xx(j)","xx(j+1)");
    /* perform test */
    for (i=1;i<=19;i++) {
        x = -100.0+200.0*i/20.0;
        locate(xx,N,x,&j);
        if ((j < N) && (j > 0))
            printf("%10.4f %6d %12.6f %12.6f\n",
                x,j,xx[j],xx[j+1]);
        else if (j == N)
            printf("%10.4f %6d %12.6f %s\n",
                x,j,xx[j],"    upper lim");
        else
            printf("%10.4f %6d %s %12.6f \n",
                x,j,"    lower lim",xx[j+1]);
    }
    free_vector(xx,1,N);
}
```

Routine hunt serves the same function as locate, but is used when the table is to be searched many times and the abscissa each time is close to its value on the previous search. xhunt.c sets up the array xx[i] and then a series x of points to locate. The hunt begins with a trial value ji (which is fed to hunt through variable j) and hunt returns solution j such that x lies between xx[j] and xx[j+1]. The two cases j=0 and j=n have the same meaning as in xlocate.c and are treated in the same way.

```
/* Driver for routine HUNT */

#include <stdio.h>
#include <math.h>
#include "nr.h"
#include "nrutil.h"

#define N 100
```

```
main()
{
    int i,j,ji;
    float x,*xx;

    xx=vector(1,N);
    /* create array to be searched */
    for (i=1;i<=N;i++)
        xx[i]=exp(i/20.0)-74.0;
    printf("\n  result of:   j=0 indicates x too small\n");
    printf("%14s j=100 indicates x too large"," ");
    printf("\n%12s %8s %4s %11s %13s \n",
        "locate:","guess","j","xx(j)","xx(j+1)");
    /* do test */
    for (i=1;i<=19;i++) {
        x = -100.0+10.0*i;
        /* trial parameter */
        j=(ji=5*i);
        /* begin search */
        hunt(xx,N,x,&j);
        if ((j < N) && (j > 0))
            printf("%12.5f %6d %6d %12.6f %12.6f \n",
                x,ji,j,xx[j],xx[j+1]);
        else if (j == N)
            printf("%12.5f %6d %6d %12.6f %s \n",
                x,ji,j,xx[j],"   upper lim");
        else
            printf("%12.5f %6d %6d %s %12.6f \n",
                x,ji,j,"   lower lim",xx[j+1]);
    }
    free_vector(xx,1,N);
}
```

The next two demonstration programs, xpolcoe.c and xpolcof.c, are so nearly identical that they may be discussed together. polcoe and polcof themselves both find coefficients of interpolating polynomials. In the present instance we have tried both a sine function and an exponential function for ya[i], each tabulated at uniformly spaced points xa[i]. The validity of the array of polynomial coefficients coeff is tested by calculating the value sum of the polynomials at a series of test points and listing these alongside the functions f which they represent.

```
/* Driver for routine POLCOE */

#include <stdio.h>
#include <math.h>
#include "nr.h"
#include "nrutil.h"

#define NP 4
#define PI 3.1415926

main()
{
    int i,j,nfunc;
    float f,sum,x,*coeff,*xa,*ya;

    coeff=vector(0,NP);
```

```
        xa=vector(0,NP);
        ya=vector(0,NP);
        for (nfunc=1;nfunc<=2;nfunc++) {
            if (nfunc == 1) {
                printf("sine function from 0 to PI\n\n");
                for (i=0;i<=NP;i++) {
                    xa[i]=(i+1)*PI/(NP+1);
                    ya[i]=sin(xa[i]);
                }
            } else if (nfunc == 2) {
                printf("exponential function from 0 to 1\n\n");
                for (i=0;i<=NP;i++) {
                    xa[i]=1.0*(i+1)/(NP+1);
                    ya[i]=exp(xa[i]);
                }
            } else {
                free_vector(ya,0,NP);
                free_vector(xa,0,NP);
                free_vector(coeff,0,NP);
                return;
            }
            polcoe(xa,ya,NP,coeff);
            printf("  coefficients\n");
            for (i=0;i<=NP;i++) printf("%12.6f",coeff[i]);
            printf("\n\n%9s %13s %15s\n","x","f(x)","polynomial");
            for (i=1;i<=10;i++) {
                if (nfunc == 1) {
                    x=(-0.05+i/10.0)*PI;
                    f=sin(x);
                } else if (nfunc == 2) {
                    x = -0.05+i/10.0;
                    f=exp(x);
                }
                sum=coeff[NP];
                for (j=NP-1;j>=0;j--)
                    sum=coeff[j]+sum*x;
                printf("%12.6f %12.6f %12.6f\n",x,f,sum);
            }
            printf("\n************************************\n");
            printf("press RETURN\n");
            getchar();
        }
}
/* Driver for routine POLCOF */

#include <stdio.h>
#include <math.h>
#include "nr.h"
#include "nrutil.h"

#define NP 4
#define PI 3.1415926

main()
{
    int i,j,nfunc;
```

```
    float f,sum,x,*coeff,*xa,*ya;

    coeff=vector(0,NP);
    xa=vector(0,NP);
    ya=vector(0,NP);
    for (nfunc=1;nfunc<=2;nfunc++) {
        if (nfunc == 1) {
            printf("sine function from 0 to PI\n\n");
            for (i=0;i<=NP;i++) {
                xa[i]=(i+1)*PI/(NP+1);
                ya[i]=sin(xa[i]);
            }
        } else if (nfunc == 2) {
            printf("exponential function from 0 to 1\n\n");
            for (i=0;i<=NP;i++) {
                xa[i]=1.0*(i+1)/(NP+1);
                ya[i]=exp(xa[i]);
            }
        } else {
            free_vector(ya,0,NP);
            free_vector(xa,0,NP);
            free_vector(coeff,0,NP);
            return;
        }
        polcof(xa,ya,NP,coeff);
        printf("  coefficients\n");
        for (i=0;i<=NP;i++) printf("%12.6f",coeff[i]);
        printf("\n\n%9s %13s %15s\n","x","f(x)","polynomial");
        for (i=1;i<=10;i++) {
            if (nfunc == 1) {
                x=(-0.05+i/10.0)*PI;
                f=sin(x);
            } else if (nfunc == 2) {
                x = -0.05+i/10.0;
                f=exp(x);
            }
            sum=coeff[NP];
            for (j=NP-1;j>=0;j--)
                sum=coeff[j]+sum*x,
            printf("%12.6f %12.6f %12.6f\n",x,f,sum);
        }
        printf("\n**********************************\n");
        printf("press RETURN\n");
        getchar();
    }
}
```

For two-dimensional interpolation, polin2 implements a bilinear interpolation. We feed it coordinates x1a,x2a for an $M \times N$ array of gridpoints as well as the function value at each gridpoint. In return it gives the value y of the interpolated function at a given point x1,x2, and the estimated accuracy dy of the interpolation. xpolin2.c runs the test on a uniform grid for the function $f(x,y) = \sin x \exp y$. Then, for an offset grid of test points, the interpolated value y is compared to the actual function value f, and the actual error is compared to the estimated error dy.

```
/* Driver for routine POLIN2 */

#include <stdio.h>
#include <math.h>
#include "nr.h"
#include "nrutil.h"

#define N 5
#define PI 3.1415926

main()
{
    int i,j;
    float dy,f,x1,x2,y,*x1a,*x2a,**ya;

    x1a=vector(1,N);
    x2a=vector(1,N);
    ya=matrix(1,N,1,N);
    for (i=1;i<=N;i++) {
        x1a[i]=i*PI/N;
        for (j=1;j<=N;j++) {
            x2a[j]=1.0*j/N;
            ya[i][j]=sin(x1a[i])*exp(x2a[j]);
        }
    }
    /* test 2-dimensional interpolation */
    printf("\nTwo dimensional interpolation of sin(x1)exp(x2)\n");
    printf("%9s %12s %13s %16s %11s\n",
        "x1","x2","f(x)","interpolated","error");
    for (i=1;i<=4;i++) {
        x1=(-0.1+i/5.0)*PI;
        for (j=1;j<=4;j++) {
            x2 = -0.1+j/5.0;
            f=sin(x1)*exp(x2);
            polin2(x1a,x2a,ya,N,N,x1,x2,&y,&dy);
            printf("%12.6f %12.6f %12.6f %12.6f %15.6f\n",
                x1,x2,f,y,dy);
        }
        printf ("***********************************\n");
    }
    free_matrix(ya,1,N,1,N);
    free_vector(x2a,1,N);
    free_vector(x1a,1,N);
}
```

Bicubic interpolation in two dimensions is carried out with bcucof and bcuint. The first supplies interpolating coefficients within a grid square and the second calculates interpolated values. The calculation provides not only interpolated function values, but also interpolated values of two partial derivatives, all of which are guaranteed to be smooth. To get this, we are required to supply more information than we have needed in previous interpolation routines.

Demonstration program xbcucof.c works with the function $f(x,y) = xy \exp(-xy)$. You may compare the two first derivatives and the cross derivative of this function with what you find computed in the routine. The function and derivatives are calculated at the four corners of a rectangular grid cell, in this case a 2×2 unit square with one corner at the origin. The points are supplied counterclockwise around the cell. d1 and d2 are the dimensions of the cell. A call to bcucof provides sixteen coefficients which are listed below for your reference.

```
Coefficients for bicubic interpolation
    0.000000E+00    0.000000E+00    0.000000E+00    0.000000E+00
    0.000000E+00    0.400000E+01    0.000000E+00    0.000000E+00
    0.000000E+00    0.000000E+00   -0.136556E+02    0.609517E+01
    0.000000E+00    0.000000E+00    0.609517E+01   -0.246149E+01
```

```c
/* Driver for routine BCUCOF */

#include <stdio.h>
#include <math.h>
#include "nr.h"
#include "nrutil.h"

main()
{
    int i,j;
    float d1,d2,ee,x1x2;
    float y[5],y1[5],y2[5],y12[5],**c;
    static float x1[]={0.0,0.0,2.0,2.0,0.0};
    static float x2[]={0.0,0.0,0.0,2.0,2.0};

    c=matrix(1,4,1,4);
    d1=x1[2]-x1[1];
    d2=x2[4]-x2[1];
    for (i=1;i<=4;i++) {
        x1x2=x1[i]*x2[i];
        ee=exp(-x1x2);
        y[i]=x1x2*ee;
        y1[i]=x2[i]*(1.0-x1x2)*ee;
        y2[i]=x1[i]*(1.0-x1x2)*ee;
        y12[i]=(1.0-3.0*x1x2+x1x2*x1x2)*ee;
    }
    bcucof(y,y1,y2,y12,d1,d2,c);
    printf("\nCoefficients for bicubic interpolation:\n\n");
    for (i=1;i<=4;i++) {
        for (j=1;j<=4;j++) printf("%12.6f",c[i][j]);
        printf("\n");
    }
    free_matrix(c,1,4,1,4);
}
```

Program xbcuint.c works with the function $f(x,y) = (xy)^2$, which has derivatives $\partial f/\partial x = 2xy^2$, $\partial f/\partial y = 2yx^2$, and $\partial^2 f/\partial x \partial y = 4xy$. These are supplied to bcuint along with the locations of the grid points. bcuint calls bcucof internally to determine coefficients, and then calculates ansy, ansy1, ansy2, the interpolated values of f, $\partial f/\partial x$ and $\partial f/\partial y$ at the specified test point (x1, x2). These are compared by the demonstration program to expected values for the three quantities, which are called ey, ey1, and ey2. The test points run along the diagonal of the grid square.

```c
/* Driver for routine BCUINT */

#include <stdio.h>
#include "nr.h"
```

```
main()
{
    int i;
    float ansy,ansy1,ansy2,ey,ey1,ey2;
    float x1,x1l,x1u,x1x2,x2,x2l,x2u,xxyy;
    float y[5],y1[5],y12[5],y2[5];
    static float xx[]={0.0,0.0,2.0,2.0,0.0};
    static float yy[]={0.0,0.0,0.0,2.0,2.0};

    x1l=xx[1];
    x1u=xx[2];
    x2l=yy[1];
    x2u=yy[4];
    for (i=1;i<=4;i++) {
        xxyy=xx[i]*yy[i];
        y[i]=xxyy*xxyy;
        y1[i]=2.0*yy[i]*xxyy;
        y2[i]=2.0*xx[i]*xxyy;
        y12[i]=4.0*xxyy;
    }
    printf("\n%6s %8s %7s %11s %6s %10s %6s %10s \n\n",
        "x1","x2","y","expect","y1","expect","y2","expect");
    for (i=1;i<=10;i++) {
        x2=(x1=0.2*i);
        bcuint(y,y1,y2,y12,x1l,x1u,x2l,x2u,x1,x2,&ansy,&ansy1,&ansy2);
        x1x2=x1*x2;
        ey=x1x2*x1x2;
        ey1=2.0*x2*x1x2;
        ey2=2.0*x1*x1x2;
        printf("%8.4f %8.4f %8.4f %8.4f %8.4f %8.4f %8.4f %8.4f\n",
            x1,x2,ansy,ey,ansy1,ey1,ansy2,ey2);
    }
}
```

Routines `splie2` and `splin2` work as a pair to perform bicubic spline interpolations. `splie2` takes a function tabulated on an $M \times N$ grid and performs one dimensional natural cubic splines along the rows of the grid to generate an array of second derivatives. These are fodder for `splin2` which takes the grid points, function values, and second derivative values and returns the interpolated function value for a desired point in the grid region.

Demonstration program `xsplie2.c` exercises `splie2` on a regular 10×10 grid of points with coordinates x1 and x2, for the function $y = (x_1 x_2)^2$. The calculated second derivative array is compared with the actual second derivative $2x_1 x_2$ of the function. Keep in mind that a natural spline is assumed, so that agreement will not be so good near the boundaries of the grid. (This shows that you should *not* assume a natural spline if you have better derivative information at the endpoints.)

```
/* Driver for routine SPLIE2 */

#include <stdio.h>
#include "nr.h"
#include "nrutil.h"

#define M 10
```

```
#define N 10

main()
{
    int i,j;
    float x1x2,*x1,*x2,**y,**y2;

    x1=vector(1,N);
    x2=vector(1,N);
    y=matrix(1,M,1,N);
    y2=matrix(1,M,1,N);
    for (i=1;i<=M;i++) x1[i]=0.2*i;
    for (i=1;i<=N;i++) x2[i]=0.2*i;
    for (i=1;i<=M;i++)
        for (j=1;j<=N;j++) {
            x1x2=x1[i]*x2[j];
            y[i][j]=x1x2*x1x2;
        }
    splie2(x1,x2,y,M,N,y2);
    printf("\nsecond derivatives from SPLIE2\n");
    printf("natural spline assumed\n");
    for (i=1;i<=5;i++) {
        for (j=1;j<=5;j++) printf("%12.6f",y2[i][j]);
        printf("\n");
    }
    printf("\nactual second derivatives\n");
    for (i=1;i<=5;i++) {
        for (j=1;j<=5;j++) printf("%12.6f",2.0*x1[i]*x1[i]);
        printf("\n");
    }
    free_matrix(y2,1,M,1,N);
    free_matrix(y,1,M,1,N);
    free_vector(x2,1,N);
    free_vector(x1,1,N);
}
```

The demonstration program xsplin2.c establishes a similar 10×10 grid for the function $y = x_1 x_2 \exp(-x_1 x_2)$. It makes a single call to splie2 to produce second derivatives y2, and then finds function values f through calls to splin2, comparing them to actual function values ff. These values are determined, for no reason better than perversity, along a quadratic path $x_2 = x_1^2$ through the grid region.

```
/* Driver for routine SPLIN2 */

#include <stdio.h>
#include <math.h>
#include "nr.h"
#include "nrutil.h"

#define M 10
#define N 10

main()
{
    int i,j;
    float f,ff,x1x2,xx1,xx2,*x1,*x2,**y,**y2;
```

```
    x1=vector(1,N);
    x2=vector(1,N);
    y=matrix(1,M,1,N);
    y2=matrix(1,M,1,N);
    for (i=1;i<=M;i++)  x1[i]=0.2*i;
    for (i=1;i<=N;i++)  x2[i]=0.2*i;
    for (i=1;i<=M;i++) {
        for (j=1;j<=N;j++) {
            x1x2=x1[i]*x2[j];
            y[i][j]=x1x2*exp(-x1x2);
        }
    }
    splie2(x1,x2,y,M,N,y2);
    printf("%9s %12s %14s %12s\n","x1","x2","splin2","actual");
    for (i=1;i<=10;i++) {
        xx1=0.1*i;
        xx2=xx1*xx1;
        splin2(x1,x2,y,y2,M,N,xx1,xx2,&f);
        x1x2=xx1*xx2;
        ff=x1x2*exp(-x1x2);
        printf("%12.6f %12.6f %12.6f %12.6f\n",xx1,xx2,f,ff);
    }
    free_matrix(y2,1,M,1,N);
    free_matrix(y,1,M,1,N);
    free_vector(x2,1,N);
    free_vector(x1,1,N);
}
```

Chapter 4: Integration of Functions

Numerical integration, or "quadrature", has been treated with some degree of detail in *Numerical Recipes* . Chapter 4 begins with `trapzd`, a procedure for applying the extended trapezoidal rule. It can be used in successive calls for sequentially improving accuracy, and is used as a foundation for several other programs. For example `qtrap` is an integrating routine that makes repeated calls to `trapzd` until a certain fractional accuracy is achieved. `qsimp` also calls `trapzd`, and in this case performs integration by Simpson's rule. Romberg integration, a generalization of Simpson's rule to successively higher orders, is performed with `qromb`—this one also calls `trapzd`. For improper integrals a different "workhorse" is used, the procedure `midpnt`. This routine applies the extended midpoint rule to avoid function evaluations at an endpoint of the region of integration. It can be used in `qtrap` or `qsimp` in place of `trapzd`. Routine `qromb` can be generalized similarly, and we have implemented this idea in `qromo`, a Romberg integrator for open intervals. The chapter also offers a number of exact replacements for `midpnt`, to be used for various types of singularity in the integrand:

1. `midinf` – if one or the other of the limits of integration is infinite.

2. `midsql` - if there is an inverse square root singularity of the integrand at the lower limit of integration.

3. `midsqu` - if there is an inverse square root singularity of the integrand at the upper limit of integration.

4. `midexp` - when the upper limit of integration is infinite and the integrand decreases exponentially at infinity.

The somewhat more subtle method of Gaussian quadrature uses unequally spaced abscissas, and weighting coefficients which can be read from tables. Routine `qgaus` computes integrals with a ten-point Gauss-Legendre weighting using such coefficients. `gauleg` calculates the tables of abscissas and weights that would apply to an N-point Gauss-Legendre quadrature.

★ ★ ★ ★

`trapzd` applies the extended trapezoidal rule for integration. It is called sequentially for higher and higher stages of refinement of the integral. The sample program `xtrapzd.c` uses `trapzd` to perform a numerical integration of the function

$$\texttt{func} = x^2(x^2 - 2)\sin x$$

whose definite integral is

$$\texttt{fint} = 4x(x^2 - 7)\sin x - (x^4 - 14x^2 + 28)\cos x.$$

The integral is performed from $A = 0.0$ to $B = \pi/2$. To demonstrate the increasing accuracy on sequential calls, $\texttt{trapzd}$ is called 12 times with the index $\texttt{i}$ increasing by one each time. The improving values of the integral are listed for comparison to the actual value $\texttt{fint}(B) - \texttt{fint}(A)$.

```
/* Driver for routine TRAPZD */

#include <stdio.h>
#include <math.h>
#include "nr.h"

#define NMAX 12
#define PIO2 1.5707963

/* Test function */
float func(x)
float x;
{
    return (x*x)*(x*x-2.0)*sin(x);
}

/* Integral of test function */
float fint(x)
float x;
{
    return 4.0*x*(x*x-7.0)*sin(x)-
        (pow(x,4.0)-14.0*(x*x)+28.0)*cos(x);
}

main()
{
    int i;
    float a=0.0,b=PIO2,s;

    printf("\nIntegral of func with 2^(n-1) points\n");
    printf("Actual value of integral is %10.6f\n",fint(b)-fint(a));
    printf("%6s %24s\n","n","approx. integral");
    for (i=1;i<=NMAX;i++) {
        s=trapzd(func,a,b,i);
        printf("%6d %20.6f\n",i,s);
    }
}
```

$\texttt{qtrap}$ carries out the same integration algorithm but allows us to specify the accuracy with which we wish the integration done. (It is specified within $\texttt{qtrap}$ as $\texttt{EPS}=1.0\texttt{e}-5$.) $\texttt{qtrap}$ itself makes the sequential calls to $\texttt{trapzd}$ until the desired accuracy is reached. Then $\texttt{qtrap}$ issues a single result. In sample program $\texttt{xqtrap.c}$ we compare this result to the exact value of the integral.

```
/* Driver for routine QTRAP */

#include <stdio.h>
#include <math.h>
```

```
#include "nr.h"

#define PIO2 1.5707963

/* Test function */
float func(x)
float x;
{
    return x*x*(x*x-2.0)*sin(x);
}

/* Integral of test function */
float fint(x)
float x;
{
    return 4.0*x*(x*x-7.0)*sin(x)-
        (pow(x,4.0)-14.0*x*x+28.0)*cos(x);
}

main()
{
    float a=0.0,b=PIO2,s;

    printf("Integral of func computed with QTRAP\n\n");
    printf("Actual value of integral is %12.6f\n",fint(b)-fint(a));
    s=qtrap(func,a,b);
    printf("Result from routine QTRAP is %12.6f\n",s);
}
```

Alternatively, the integral may be handled by qsimp which applies Simpson's rule. Sample program xqsimp.c carries out the same integration as the previous program, and reports the result in the same way as well.

```
/* Driver for routine QSIMP */

#include <stdio.h>
#include <math.h>
#include "nr.h"

#define PIO2 1.5707963

/* Test function */
float func(x)
float x;
{
    return x*x*(x*x-2.0)*sin(x);
}

/* Integral of test function */
float fint(x)
float x;
{
    return 4.0*x*(x*x-7.0)*sin(x)-
        (pow(x,4.0)-14.0*x*x+28.0)*cos(x);
}

main()
```

```
{
    float a=0.0,b=PIO2,s;

    printf("Integral of func computed with QSIMP\n\n");
    printf("Actual value of integral is %12.6f\n",fint(b)-fint(a));
    s=qsimp(func,a,b);
    printf("Result from routine QSIMP is %11.6f\n",s);
}
```

qromb generalizes Simpson's rule to higher orders. It makes successive calls to trapzd and stores the results. Then it uses polint, the polynomial interpolator/extrapolator, to project the value of integral which would be obtained were we to continue indefinitely with trapzd. Sample program xqromb.c is essentially identical to the sample programs for qtrap and qsimp.

```
/* Driver for routine QROMB */

#include <stdio.h>
#include <math.h>
#include "nr.h"

#define PIO2 1.5707963

/* Test function */
float func(x)
float x;
{
    return x*x*(x*x-2.0)*sin(x);
}

/* Integral of test function func */
float fint(x)
float x;
{
    return 4.0*x*(x*x-7.0)*sin(x)
        -(pow(x,4.0)-14.0*x*x+28.0)*cos(x);
}

main()
{
    float a=0.0,b=PIO2,s;

    printf("Integral of func computed with QROMB\n\n");
    printf("Actual value of integral is %12.6f\n",fint(b)-fint(a));
    s=qromb(func,a,b);
    printf("Result from routine QROMB is %11.6f\n",s);
}
```

Sample program xmidpnt.c uses the function $func = 1/\sqrt{x}$ which is singular at the origin. Limits of integration are set at $A = 0.0$ and $B = 1.0$. midpnt, however, implements an open formula and does not evaluate the function exactly at $x = 0$. In this case the integral is compared to $fint(B) - fint(A)$ where $fint = 2\sqrt{x}$, the integral of func.

```
/* Driver for routine MIDPNT */

#include <stdio.h>
```

```
#include <math.h>
#include "nr.h"

#define NMAX 10

/* Test function */
float func(x)
float x;
{
    return 1.0/sqrt(x);
}

/* Integral of test function */
float fint(x)
float x;
{
    return 2.0*sqrt(x);
}

main()
{
    float a=0.0,b=1.0,s;
    int i;

    printf("\nIntegral of func computed with MIDPNT\n");
    printf("Actual value of integral is %7.4f\n",(fint(b)-fint(a)));
    printf("%6s %29s \n","n","Approx. integral");
    for (i=1;i<=NMAX;i++) {
        s=midpnt(func,a,b,i);
        printf("%6d %24.6f\n",i,s);
    }
}
```

One of the special forms of midpnt, namely midsql, is demonstrated by sample program xqromo.c. We evaluate the integral of $\sqrt{x}/\sin x$ from 0.0 to $\pi/2$. This has a $1/\sqrt{x}$ singularity at $x = 0$.

```
/* Driver for routine QROMO */

#include <stdio.h>
#include <math.h>
#include "nr.h"

#define PIO2 1.5707963

static float func(x)
float x;
{
    return (float) (sqrt(x)/sin(x));
}

main()
{
    float a=0.0,b=PIO2,result;

    printf("Improper integral:\n\n");
    result=qromo(func,a,b,midsql);
```

```
        printf("Function: sqrt(x)/sin(x)      Interval: (0,pi/2)\n");
        printf("Using: MIDSQL                 Result: %8.4f\n",result);
}
```

Procedure `qgaus` performs a Gauss-Legendre integration, using only ten function evaluations. Sample program `xqgaus.c` applies it to the function $x\exp(-x)$ whose integral from x_1 to x is $(1+x_1)\exp(-x_1) - (1+x)\exp(-x)$. `qgaus` returns the value of this integral. The method is used for a series of intervals, as short as $(0.0 - 0.5)$ and as long as $(0.0 - 5.0)$. You may observe how the accuracy depends on the interval.

```
/* Driver for routine QGAUS */

#include <stdio.h>
#include <math.h>
#include "nr.h"

#define X1 0.0
#define X2 5.0
#define NVAL 10

float func(x)
float x;
{
    return x*exp(-x);
}

main()
{
    float dx,ss,x;
    int i;

    dx=(X2-X1)/NVAL;
    printf("\n%s %10s %13s\n\n","0.0 to","qgaus","expected");
    for (i=1;i<=NVAL;i++) {
        x=X1+i*dx;
        ss=qgaus(func,X1,x);
        printf("%5.2f %12.6f %12.6f\n",x,ss,
            (-(1.0+x)*exp(-x)+(1.0+X1)*exp(-X1)));
    }
}
```

Sample program `xgauleg.c`, which drives `gauleg`, performs the same method of quadrature, and on the same function. However, it chooses its own abscissas and weights for the Gauss-Legendre calculation, and is not restricted to a ten-point formula; it can do an N-point calculation for any N. The N abscissas and weights appropriate to an interval $x = 0.0$ to 1.0 are found by sample program `xgauleg.c` for the case $N = 10$. The results you should find are listed below. Next the program applies these values to a quadrature and compares the result to that from a formal integration.

position	weight
.013047	.033336
.067468	.074726
.160295	.109543
.283302	.134633
.425563	.147762
.574437	.147762
.716698	.134633

```
        .839705          .109543
        .932532          .074726
        .986953          .033336
```

```
/* Driver for routine GAULEG */

#include <stdio.h>
#include <math.h>
#include "nr.h"
#include "nrutil.h"

#define NPOINT 10
#define X1 0.0
#define X2 1.0
#define X3 10.0

float func(x)
float x;
{
    return x*exp(-x);
}

main()
{
    int i;
    float xx;
    double *x,*w;

    x=dvector(1,NPOINT);
    w=dvector(1,NPOINT);
    gauleg(X1,X2,x,w,NPOINT);
    printf("\n%2s %10s %12s\n","#","x[i]","w[i]");
    for (i=1;i<=NPOINT;i++)
        printf("%2d %12.6f %12.6f\n",i,x[i],w[i]);
    /* Demonstrate the use of gauleg for integration */
    gauleg(X1,X3,x,w,NPOINT);
    xx=0.0;
    for (i=1;i<=NPOINT;i++)
        xx += (w[i]*func(x[i]));
    printf("\nIntegral from GAULEG: %12.6f\n",xx);
    printf("Actual value: %12.6f\n",
        (1.0+X1)*exp(-X1)-(1.0+X3)*exp(-X3));
    free_dvector(w,1,NPOINT);
    free_dvector(x,1,NPOINT);
}
```

Chapter 4 of *Numerical Recipes* ends with a short discussion of multidimensional integration, exemplified by routine quad3d which does a 3-dimensional integration by repeated 1-dimensional integration. The C version of this algorithm is quite simple because recursion can be used. Sample program xquad3d.c applies the method to the integration of func $= x^2 + y^2 + z^2$ over a spherical volume with a radius xmax which is taken successively as $0.1, 0.2, ..., 1.0$. The integral is done in Cartesian rather than spherical coordinates, but the result is compared to that easily found in spherical coordinates, $4\pi(\text{xmax})^5/5$. Procedure func generates the function. Procedures yy1 and yy2 supply the two limits of the y-integration for each value of x. Similarly z1 and z2 give the limits of z-integration for given x and y.

```c
/* Driver for routine QUAD3D */

#include <stdio.h>
#include <math.h>
#include "nr.h"

#define PI 3.1415927
#define NVAL 10

static float xmax;

float func(x,y,z)
float x,y,z;
{
    return x*x+y*y+z*z;
}

float z1(x,y)
float x,y;
{
    return (float) -sqrt(xmax*xmax-x*x-y*y);
}

float z2(x,y)
float x,y;
{
    return (float) sqrt(xmax*xmax-x*x-y*y);
}

float yy1(x)
float x;
{
    return (float) -sqrt(xmax*xmax-x*x);
}

float yy2(x)
float x;
{
    return (float) sqrt(xmax*xmax-x*x);
}

main()
{
    int i;
    float xmin,s;

    printf("Integral of r^2 over a spherical volume\n\n");
    printf("%13s %10s %11s\n","radius","QUAD3D","Actual");
    for (i=1;i<=NVAL;i++) {
        xmax=0.1*i;
        xmin = -xmax;
        s=quad3d(func,xmin,xmax);
        printf("%12.2f %12.6f %11.6f\n",
            xmax,s,4.0*PI*pow(xmax,5.0)/5.0);
    }
}
```

Chapter 5: Evaluation of Functions

Chapter 5 of *Numerical Recipes* treats the approximation and evaluation of functions. The methods, along with a few others, are applied in Chapter 6 to the calculation of a collection of "special" functions. Polynomial or power series expansions are perhaps the most often used approximations and a few tips are given for accelerating the convergence of some series. In the case of alternating series, Euler's transformation is popular, and is implemented in program `eulsum`. For general polynomials, `ddpoly` demonstrates the evaluation of both the polynomial and its derivatives from a list of its coefficients. The division of one polynomial into another, giving a quotient and remainder polynomial, is done by `poldiv`.

The approximation of functions by Chebyshev polynomial series is presented as a method of arriving at the approximation of nearly smallest deviation from the true function over a given region for a specified order of approximation. The coefficients for such polynomials are given by `chebft` and function approximations are subsequently carried out by `chebev`. To generate the derivative or integral of a function from its Chebyshev coefficients, use `chder` or `chint` respectively. Finally, to convert Chebyshev coefficients into coefficients of a polynomial for the same function (a dangerous procedure about which we offer due warning in the text) use `chebpc` and `pcshft` in succession.

Chapter 5 also treats several methods for which we supply no programs. These are continued fractions, rational functions, recurrence relations, and the solution of quadratic and cubic equations.

$$\star \quad \star \quad \star \quad \star$$

Procedure `eulsum` applies Euler's transformation to the summation of an alternating series. It is called successively for each term to be summed. Our sample program `xeulsum.c` evaluates the approximation

$$\ln(1 + x) = x - \frac{x^2}{2} + \frac{x^3}{3} - \frac{x^4}{4} + \cdots \qquad -1 < x < 1$$

It asks how many terms `mval` are to be included in the approximation and then makes `mval` calls to `eulsum`. Each time, index j increases and `term` takes the value $(-1)^{j+1}x^j/j$. Both this approximation and the function $\ln(1 + x)$ itself are evaluated across the region -1 to 1 for comparison. If `mval` is set less than 1 or more than 40, the program terminates.

```
/* Driver for routine EULSUM */

#include <stdio.h>
#include <math.h>
#include "nr.h"
#include "nrutil.h"

#define NVAL 40

main()
{
    int i,j,mval;
    float sum,term,x,xpower,*wksp;

    wksp=vector(1,NVAL);
    /* evaluate ln(1+x)=x-x^2/2+x^3/3-x^4/4 ... for -1<x<1 */
    for (;;) {
        printf("\nHow many terms in polynomial?\n");
        printf("Enter n between 1 and %2d. (n=0 to end) ",NVAL);
        scanf("%d",&mval);
        printf("\n");
        if ((mval <= 0) || (mval > NVAL)) {
            free_vector(wksp,1,NVAL);
            return;
        }
        printf("%9s %14s %14s\n","x","actual","polynomial");
        for (i = -8;i<=8;i++) {
            x=i/10.0;
            sum=0.0;
            xpower = -1;
            for (j=1;j<=mval;j++) {
                xpower *= (-x);
                term=xpower/j;
                eulsum(&sum,term,j,wksp);
            }
            printf("%12.6f %12.6f %12.6f\n",x,log(1.0+x),sum);
        }
    }
}
```

ddpoly evaluates a polynomial and its derivatives, given the coefficients of the polynomial in the form of an input vector. Sample program xddpoly.c illustrates this for the polynomial:

$$(x-1)^5 = -1 + 5x - 10x^2 + 10x^3 - 5x^4 + x^5$$

(This is a foolish example, of course. No one would knowingly evaluate $(x-1)^5$ by multiplying it out and evaluating terms individually—but it gives us a convenient way to check the result!). Since this is a fifth order polynomial, we set NC, the degree of the polynomial, to 5, and initialize the array c of coefficients, with c[0] being the constant coefficient and c[5] the highest-order coefficient. There are two loops, one of which evaluates for x values from 0.0 to 2.0, and the other of which stores the value of the function and NC-1 derivatives. d[j][i] keeps the entire array of values for printing. In the second part of the program, the polynomial evaluations are compared with

$$f^{(n-1)}(x) = \frac{5!}{(6-n)!}(x-1.0)^{6-n} \qquad n = 1, \ldots, 5$$

```
/* Driver for routine DDPOLY */

#include <stdio.h>
#include "nr.h"
#include "nrutil.h"

#define NC 5
#define ND NC-1
#define NP 20

main()
{
    int i,j,k;
    float x,pwr,*pd,**d;
    static float c[NC+1]={-1.0,5.0,-10.0,10.0,-5.0,1.0};
        static char *a[ND+1]={"polynomial:", "first deriv:",
        "second deriv:","third deriv:","fourth deriv:"};

    pd=vector(0,ND);
    d=matrix(0,ND,1,NP);
    for (i=1;i<=NP;i++) {
        x=0.1*i;
        ddpoly(c,NC,x,pd,ND);
        for (j=0;j<=ND;j++) d[j][i]=pd[j];
    }
    for (i=0;i<=ND;i++) {
        printf("%6s %s \n"," ",a[i]);
        printf("%12s %17s %15s\n","x","DDPOLY","actual");
        for (j=1;j<=NP;j++) {
            x=0.1*j;
            pwr=1.0;
            for (k=1;k<=NC-i;k++) pwr *= x-1.0;
            printf("%15.6f %15.6f %15.6f\n",x,d[i][j],
                (factrl(NC)/factrl(NC-i))*pwr);
        }
        printf("press ENTER to continue...\n");
        getchar();
    }
    free_matrix(d,1,ND,1,NP);
    free_vector(pd,0,ND);
}
```

poldiv divides polynomials. Given the coefficients of a numerator and denominator polynomial, poldiv returns the coefficients of a quotient and a remainder polynomial. Sample program xpoldiv.c takes

$$\text{Numerator} = u = -1 + 5x - 10x^2 + 10x^3 - 5x^4 + x^5 = (x-1)^5$$

$$\text{Denominator} = v = 1 + 3x + 3x^2 + x^3 = (x+1)^3$$

for which we expect

$$\text{Quotient} = q = 31 - 8x + x^2$$

$$\text{Remainder} = r = -32 - 80x - 80x^2$$

The program compares these with the output of poldiv.

```
/* Driver for routine POLDIV */

#include <stdio.h>
#include "nr.h"
#include "nrutil.h"

#define N 5
#define NV 3

main()
{
    int i;
    static float u[N+1]={-1.0,5.0,-10.0,10.0,-5.0,1.0};
    static float v[NV+1]={1.0,3.0,3.0,1.0};
    float *q,*r;

    q=vector(0,N);
    r=vector(0,N);
    poldiv(u,N,v,NV,q,r);
    printf("\n%10s %10s %10s %10s %10s %10s\n\n",
        "x^0","x^1","x^2","x^3","x^4","x^5");
    printf("quotient polynomial coefficients:\n");
    for (i=0;i<=5;i++) printf("%10.2f ",q[i]);
    printf("\nexpected quotient coefficients:\n");
    printf("%10.2f %10.2f %10.2f %10.2f %10.2f %10.2f\n\n",
        31.0,-8.0,1.0,0.0,0.0,0.0);
    printf("remainder polynomial coefficients:\n");
    for (i=0;i<=3;i++) printf("%10.2f ",r[i]);
    printf("\nexpected remainder coefficients:\n");
    printf("%10.2f %10.2f %10.2f %10.2f\n",-32.0,-80.0,-80.0,0.0);
    free_vector(r,0,N);
    free_vector(q,0,N);
}
```

The remaining six programs all deal with Chebyshev polynomials. chebft evaluates the coefficients for a Chebyshev polynomial approximation of a function on a specified interval and for a maximum degree N of polynomial. Demonstration program xchebft.c uses the function func $= x^2(x^2-2)\sin x$ on the interval $(-\pi/2, \pi/2)$ with the maximum degree of NVAL=40. Notice that chebft is called with this maximum degree specified, even though subsequent evaluations may truncate the Chebyshev series at much lower terms. After we choose the number mval of terms in the evaluation, the Chebyshev polynomial is evaluated term by term, for x values between -0.8π and 0.8π, and the result f is compared to the actual function value.

```
/* Driver for routine CHEBFT */

#include <stdio.h>
#include <math.h>
#include "nr.h"

#define NVAL 40
#define PIO2 1.5707963

float func(x)
float x;
{
```

```
        return x*x*(x*x-2.0)*sin(x);
}

main()
{
    float a=(-PIO2),b=PIO2,dum,f;
    float t0,t1,term,x,y,c[NVAL];
    int i,j,mval;

    chebft(a,b,c,NVAL,func);
    /* test result */
    for (;;) {
        printf("\nHow many terms in Chebyshev evaluation?\n");
        printf("Enter n between 6 and %2d. (n=0 to end).\n",NVAL);
        scanf("%d",&mval);
        if ((mval <= 0) || (mval > NVAL)) break;
        printf("\n%9s %14s %16s\n","x","actual","chebyshev fit");
        for (i = -8;i<=8;i++) {
            x=i*PIO2/10.0;
            y=(x-0.5*(b+a))/(0.5*(b-a));
            /* Evaluate Chebyshev polynomial without CHEBEV */
            t0=1.0;
            t1=y;
            f=c[1]*t1+c[0]*0.5;
            for (j=2;j<mval;j++) {
                dum=t1;
                t1=2.0*y*t1-t0;
                t0=dum;
                term=c[j]*t1;
                f += term;
            }
            printf("%12.6f %12.6f %12.6f\n",x,func(x),f);
        }
    }
}
```

chebev is the Chebyshev polynomial evaluator, and the next sample program xchebev.c uses it for the same problem just discussed. In fact, the program is identical except that it replaces the internal polynomial summation with chebev, which applies Clenshaw's recurrence to find the polynomial values.

```
/* Driver for routine CHEBEV */

#include <stdio.h>
#include <math.h>
#include "nr.h"

#define NVAL 40
#define PIO2 1.5707963

float func(x)
float x;
{
    return x*x*(x*x-2.0)*sin(x);
}

main()
```

```
{
    int i,mval;
    float a=(-PIO2),b=PIO2,x,c[NVAL];

    chebft(a,b,c,NVAL,func);
    /* Test Chebyshev evaluation routine */
    for (;;) {
        printf("\nHow many terms in Chebyshev evaluation?\n");
        printf("Enter n between 6 and %2d. (n=0 to end).\n",NVAL);
        scanf("%d",&mval);
        if ((mval <= 0) || (mval > NVAL)) break;
        printf("\n%9s %14s %16s \n","x","actual","chebyshev fit");
        for (i = -8;i<=8;i++) {
            x=i*PIO2/10.0;
            printf("%12.6f %12.6f %12.6f\n",
                x,func(x),chebev(a,b,c,mval,x));
        }
    }
}
```

By the same token, the tests for `chint` and `chder` needn't be much different. `chint` determines Chebyshev coefficients for the integral of the function, and `chder` for the derivative of the function, given the Chebyshev coefficients for the function itself (from `chebft`) and the interval (A, B) of evaluation. When applied to the function above, the true integral is

$$\text{fint} = 4x(x^2 - 7)\sin x - (x^4 - 14x^2 + 28)\cos x$$

and the true derivative is

$$\text{fder} = 4x(x^2 - 1)\sin x + x^2(x^2 - 2)\cos x$$

The code in sample programs `xchint.c` and `xchder.c` compares the true and Chebyshev-derived integral and derivative values for a range of x in the interval of evaluation. Since `chint` and `chder` return Chebyshev coefficients, and not the integral and derivative values themselves, calls to `chebev` are required for the comparison.

```
/* Driver for routine CHINT */

#include <stdio.h>
#include <math.h>
#include "nr.h"

#define NVAL 40
#define PIO2 1.5707963

float func(x)
float x;
{
    return x*x*(x*x-2.0)*sin(x);
}

float fint(x)
float x;
{
    return 4.0*x*(x*x-7.0)*sin(x)
```

```
            -(x*x*(x*x-14.0)+28.0)*cos(x);
}

main()
{
    int i,mval;
    float a=(-PIO2),b=PIO2,x;
    float c[NVAL],cint[NVAL];

    chebft(a,b,c,NVAL,func);
    /* test integral */
    for (;;) {
        printf("\nHow many terms in Chebyshev evaluation?\n");
        printf("Enter n between 6 and %2d. (n=0 to end).\n",NVAL);
        scanf("%d",&mval);
        if ((mval <= 0) || (mval > NVAL)) break;
        chint(a,b,c,cint,mval);
        printf("\n%9s %14s %16s\n","x","actual","Cheby. integ.");
        for (i = -8;i<=8;i++) {
            x=i*PIO2/10.0;
            printf("%12.6f %12.6f %12.6f\n",
                x,fint(x)-fint(-PIO2),chebev(a,b,cint,mval,x));
        }
    }
}

/* Driver for routine CHDER */

#include <stdio.h>
#include <math.h>
#include "nr.h"

#define NVAL 40
#define PIO2 1.5707963

float func(x)
float x;
{
    return x*x*(x*x-2.0)*sin(x);
}

float fder(x)
float x;
{
    return 4.0*x*(x*x-1.0)*sin(x)+x*x*(x*x-2.0)*cos(x);
}

main()
{
    int i,mval;
    float a=(-PIO2),b=PIO2,x;
    float c[NVAL],cder[NVAL];

    chebft(a,b,c,NVAL,func);
    /* Test derivative */
    for (;;) {
        printf("\nHow many terms in Chebyshev evaluation?\n");
```

```
        printf("Enter n between 6 and %2d.  (n=0 to end).\n",NVAL);
        scanf("%d",&mval);
        if ((mval <= 0) || (mval > NVAL)) break;
        chder(a,b,c,cder,mval);
        printf("\n%9s %14s %16s\n","x","actual","Cheby. deriv.");
        for (i = -8;i<=8;i++) {
            x=i*PIO2/10.0;
            printf("%12.6f %12.6f %12.6f\n",
                x,fder(x),chebev(a,b,cder,mval,x));
        }
    }
}
```

The final two programs of this chapter turn the coefficients of a Chebyshev approximation into those of a polynomial approximation in the variable

$$y = \frac{x - \frac{1}{2}(B + A)}{\frac{1}{2}(B - A)}$$

(routine chebpc), or of a polynomial approximation in x itself (routine chebpc followed by pcshft). These procedures are discouraged for reasons discussed in *Numerical Recipes*, but should they serve some special purpose for you, we have at least warned that you will be sacrificing accuracy, particularly for polynomials above order 7 or 8. Sample program xchebpc.c calls chebft and chebpc to find polynomial coefficients in y for a truncated series. For a set of x values between $-\pi$ and π it calculates y and then the terms of the y-polynomial, which are summed in variable poly. Finally, poly is compared to the true function value. (The function func is the same as used before.)

```
/* Driver for routine CHEBPC */

#include <stdio.h>
#include <math.h>
#include "nr.h"

#define NVAL 40
#define PIO2 1.5707963

float func(x)
float x;
{
    return x*x*(x*x-2.0)*sin(x);
}

main()
{
    int i,j,mval;
    float a=(-PIO2),b=PIO2,poly,x,y;
    float c[NVAL],d[NVAL];

    chebft(a,b,c,NVAL,func);
    for (;;) {
        printf("\nHow many terms in Chebyshev evaluation?\n");
        printf("Enter n between 6 and %2d.  (n=0 to end).\n",NVAL);
        scanf("%d",&mval);
        if ((mval <= 0) || (mval > NVAL)) return;
        chebpc(c,d,mval);
```

```
    /* Test polynomial */
    printf("\n%9s %14s %14s\n","x","actual","polynomial");
    for (i = -8;i<=8;i++) {
        x=i*PIO2/10.0;
        y=(x-0.5*(b+a))/(0.5*(b-a));
        poly=d[mval-1];
        for (j=mval-2;j>=0;j--) poly=poly*y+d[j];
        printf("%12.6f %12.6f %12.6f\n",x,func(x),poly);
    }
    }
}
```

pcshft shifts the polynomial to be one in variable x. Sample program xpcshft.c is like the previous program except that it follows the call to chebpc with a call to pcshft.

```
/* Driver for routine PCSHFT */

#include <stdio.h>
#include <math.h>
#include "nr.h"

#define NVAL 40
#define PIO2 1.5707963

float func(x)
float x;
{
    return x*x*(x*x-2.0)*sin(x);
}

main()
{
    int i,j,mval;
    float a=(-PIO2),b=PIO2,poly,x;
    float c[NVAL],d[NVAL];

    chebft(a,b,c,NVAL,func);
    for (;;) {
        printf("\nHow many terms in Chebyshev evaluation?\n");
        printf("Enter n between 6 and %2d. (n=0 to end).\n",NVAL);
        scanf("%d",&mval);
        if ((mval <= 0) || (mval > NVAL)) break;
        chebpc(c,d,mval);
        pcshft(a,b,d,mval);
        /* Test shifted polynomial */
        printf("\n%9s %14s %14s\n","x","actual","polynomial");
        for (i = -8;i<=8;i++) {
            x=i*PIO2/10.0;
            poly=d[mval-1];
            for (j=mval-2;j>=0;j--) poly=poly*x+d[j];
            printf("%12.6f %12.6f %12.6f\n",x,func(x),poly);
        }
    }
}
```

Chapter 6: Special Functions

This chapter on special functions provides illustrations of techniques developed in Chapter 5. At the same time, it offers routines for calculating many of the functions that arise frequently in analytical work, but which are not so common to be included, for example, as a single keystroke on your pocket calculator. In terms of demonstration programs, they represent a simple bunch. The test routines are all virtually identical, all making reference to a single file of function values called fncval.dat *which is listed in the Appendix at the end of this chapter. In this file are accurate values for the individual functions for a variety of values for each argument. We have aimed to "stress" the routines a bit by throwing in some extreme values for the arguments.*

Many of the function values came from Abramowitz and Stegun's Handbook of Mathematical Functions. Some others, however, came from our library of dusty volumes from past masters. There is an implicit danger in a comparison test like this—namely, that our source has used the same algorithms as ours to construct the tables. In that case, we test only our mutual competence at computing, not the correctness of the result. Nevertheless, there is some assurance in knowing that the values we calculate are the ones that have been used and scrutinized for many years. Moreover, the expressions for the functions themselves can be worked out in certain special or limiting cases without computer aid, and in these instances the results have proven correct.

★   ★   ★   ★

With few exceptions, the routines that follow work in this fashion:

1. Open file fncval.dat.

2. Find the appropriate data table according to its title.

3. Read the argument list for each table entry and pass them to the routine to be tested.

4. Print the arguments along with the expected and actual results.

For the routines in this list, therefore, we forego any further comment, but simply identify them by the special function that they evaluate.

Natural logarithm of the gamma function for positive arguments:

```
/* Driver for routine GAMMLN */

#include <stdio.h>
#include <math.h>
```

53

```
#include "nr.h"
#include "nrutil.h"

#define MAXSTR 80

main()
{
    char txt[MAXSTR];
    int i,nval,strncmp();
    float actual,calc,x;
    FILE *fp;

    if ((fp = fopen("fncval.dat","r")) == NULL)
        nrerror("Data file FNCVAL.DAT not found\n");
    fgets(txt,MAXSTR,fp);
    while(strncmp(txt,"Gamma Function",14)) {
        fgets(txt,MAXSTR,fp);
        if (feof(fp)) return;
    }
    fscanf(fp,"%d %*s",&nval);
    printf("\n%s\n",txt);
    printf("%10s %21s %21s\n","x","actual","gammln(x)");
    for (i=1;i<=nval;i++) {
        fscanf(fp,"%f %f",&x,&actual);
        if (x > 0.0) {
            calc=(x<1.0 ? gammln(x+1.0)-log(x) : gammln(x));
            printf("%12.2f %20.6f %20.6f\n",x,
                log(actual),calc);
        }
    }
    fclose(fp);
}
```

Factorial function *N*!:

```
/* Driver for routine FACTRL */

#include <stdio.h>
#include <math.h>
#include "nr.h"
#include "nrutil.h"

#define MAXSTR 80

main()
{
    char txt[MAXSTR];
    float actual;
    int i,n,nval,strncmp();
    FILE *fp;

    if ((fp = fopen("fncval.dat","r")) == NULL)
        nrerror("Data file FNCVAL.DAT not found\n");
    fgets(txt,MAXSTR,fp);
    while(strncmp(txt,"N-factorial",11)) {
        fgets(txt,MAXSTR,fp);
        if (feof(fp)) return;
    }
```

```
        fscanf(fp,"%d %*s",&nval);
        printf("\n%s\n",txt);
        printf("%6s %18s %20s \n","n","actual","factrl(n)");
        for (i=1;i<=nval;i++) {
            fscanf(fp,"%d %f ",&n,&actual);
            if (actual < 1.0e10)
                printf("%6d %20.0f %20.0f\n",n,actual,factrl(n));
            else
                printf("%6d %20e %20e \n",n,actual,factrl(n));
        }
        fclose(fp);
}
```

Binomial coefficients:

```
/* Driver for routine BICO */

#include <stdio.h>
#include "nr.h"
#include "nrutil.h"

#define MAXSTR 80

main()
{
    char txt[MAXSTR];
    int i,k,n,nval,strncmp();
    float binco;
    FILE *fp;

    if ((fp = fopen("fncval.dat","r")) == NULL)
        nrerror("Data file FNCVAL.DAT not found\n");
    fgets(txt,MAXSTR,fp);
    while(strncmp(txt,"Binomial Coefficients",21)) {
        fgets(txt,MAXSTR,fp);
        if (feof(fp)) return;
    }
    fscanf(fp,"%d %*s",&nval);
    printf("\n%s\n",txt);
    printf("%6s %6s %12s %12s \n","n","k","actual","bico(n,k)");
    for (i=1;i<=nval;i++) {
        fscanf(fp,"%d %d %f ",&n,&k,&binco);
        printf("%6d %6d %12.0f %12.0f \n",n,k,binco,bico(n,k));
    }
    fclose(fp);
}
```

Natural logarithm of *N*!:

```
/* Driver for routine FACTLN */

#include <stdio.h>
#include <math.h>
#include "nr.h"
#include "nrutil.h"

#define MAXSTR 80

main()
```

```
{
    char txt[MAXSTR];
    int i,n,nval,strncmp();
    float val;
    FILE *fp;

    if ((fp = fopen("fncval.dat","r")) == NULL)
        nrerror("Data file FNCVAL.DAT not found\n");
    fgets(txt,MAXSTR,fp);
    while(strncmp(txt,"N-factorial",11)) {
        fgets(txt,MAXSTR,fp);
        if (feof(fp)) return;
    }
    fscanf(fp,"%d %*s",&nval);
    printf("\nlog of n_factorial\n");
    printf("\n%6s %19s %21s\n","n","actual","factln(n)");
    for (i=1;i<=nval;i++) {
        fscanf(fp,"%d %f",&n,&val);
        printf("%6d %20.7f %20.7f\n",n,log(val),factln(n));
    }
    fclose(fp);
}
```

Beta function:

```
/* Driver for routine BETA */

#include <stdio.h>
#include "nr.h"
#include "nrutil.h"

#define MAXSTR 80

main()
{
    char txt[MAXSTR];
    int i,nval,strncmp();
    float val,w,z;
    FILE *fp;

    if ((fp = fopen("fncval.dat","r")) == NULL)
        nrerror("Data file FNCVAL.DAT not found\n");
    fgets(txt,MAXSTR,fp);
    while(strncmp(txt,"Beta Function",13)) {
        fgets(txt,MAXSTR,fp);
        if (feof(fp)) return;
    }
    fscanf(fp,"%d %*s",&nval);
    printf("\n%s\n",txt);
    printf("%5s %6s %16s %20s\n","w","z","actual","beta(w,z)");
    for (i=1;i<=nval;i++) {
        fscanf(fp,"%f %f %f",&w,&z,&val);
        printf("%6.2f %6.2f %18.6e %18.6e\n",w,z,val,beta(w,z));
    }
    fclose(fp);
}
```

Incomplete gamma function $P(a, x)$:

```
/* Driver for routine GAMMP */

#include <stdio.h>
#include "nr.h"
#include "nrutil.h"

#define MAXSTR 80

main()
{
    char txt[MAXSTR];
    int i,nval,strncmp();
    float a,val,x;
    FILE *fp;

    if ((fp = fopen("fncval.dat","r")) == NULL)
        nrerror("Data file FNCVAL.DAT not found\n");
    fgets(txt,MAXSTR,fp);
    while (strncmp(txt,"Incomplete Gamma Function",25)) {
        fgets(txt,MAXSTR,fp);
        if (feof(fp)) return;
    }
    fscanf(fp,"%d %*s",&nval);
    printf("\n%s\n",txt);
    printf("%4s %11s %14s %14s \n","a","x","actual","gammp(a,x)");
    for (i=1;i<=nval;i++) {
        fscanf(fp,"%f %f %f",&a,&x,&val);
        printf("%6.2f %12.6f %12.6f %12.6f \n",a,x,val,gammp(a,x));
    }
    fclose(fp);
}
```

Incomplete gamma function $Q(a, x) = 1 - P(a, x)$:

```
/* Driver for routine GAMMQ */

#include <stdio.h>
#include "nr.h"
#include "nrutil.h"

#define MAXSTR 80

main()
{
    char txt[MAXSTR];
    int i,nval,strncmp();
    float a,val,x;
    FILE *fp;

    if ((fp = fopen("fncval.dat","r")) == NULL)
        nrerror("Data file FNCVAL.DAT not found\n");
    fgets(txt,MAXSTR,fp);
    while(strncmp(txt,"Incomplete Gamma Function",25)) {
        fgets(txt,MAXSTR,fp);
        if (feof(fp)) return;
    }
    fscanf(fp,"%d %*s",&nval);
    printf("\n%s\n",txt);
```

```
        printf("%4s %11s %14s %14s \n","a","x","actual","gammq(a,x)");
        for (i=1;i<=nval;i++) {
            fscanf(fp,"%f %f %f",&a,&x,&val);
            printf("%6.2f %12.6f %12.6f %12.6f\n",a,x,(1.0-val),gammq(a,x));
        }
        fclose(fp);
}
```

Incomplete gamma function $P(a, x)$ evaluated from series representation:

```
/* Driver for routine GSER */

#include <stdio.h>
#include "nr.h"
#include "nrutil.h"

#define MAXSTR 80

main()
{
    char txt[MAXSTR];
    int i,nval,strncmp();
    float a,gamser,gln,val,x;
    FILE *fp;

    if ((fp = fopen("fncval.dat","r")) == NULL)
        nrerror("Data file FNCVAL.DAT not found\n");
    fgets(txt,MAXSTR,fp);
    while(strncmp(txt,"Incomplete Gamma Function",25)) {
        fgets(txt,MAXSTR,fp);
        if (feof(fp)) return;
    }
    fscanf(fp,"%d %*s",&nval);
    printf("\n%s\n",txt);
    printf("%4s %11s %14s %14s %12s %8s\n","a","x",
        "actual","gser(a,x)","gammln(a)","gln");
    for (i=1;i<=nval;i++) {
        fscanf(fp,"%f %f %f",&a,&x,&val);
        gser(&gamser,a,x,&gln);
        printf("%6.2f %12.6f %12.6f %12.6f %12.6f %12.6f\n",
            a,x,val,gamser,gammln(a),gln);
    }
    fclose(fp);
}
```

Incomplete gamma function $Q(a, x)$ evaluated by continued fraction representation:

```
/* Driver for routine GCF */

#include <stdio.h>
#include "nr.h"
#include "nrutil.h"

#define MAXSTR 80

main()
{
    char txt[MAXSTR];
    int i,nval,strncmp();
```

```
     float a,val,x,gammcf,gln;
     FILE *fp;

     if ((fp = fopen("fncval.dat","r")) == NULL)
          nrerror("Data file FNCVAL.DAT not found\n");
     fgets(txt,MAXSTR,fp);
     while(strncmp(txt,"Incomplete Gamma Function",25)) {
          fgets(txt,MAXSTR,fp);
          if (feof(fp)) return;
     }
     fscanf(fp,"%d %*s",&nval);
     printf("\n%s\n",txt);
     printf("%4s %11s %14s %13s %13s %8s\n","a","x",
          "actual","gcf(a,x)","gammln(a)","gln");
     for (i=1;i<=nval;i++) {
          fscanf(fp,"%f %f %f",&a,&x,&val);
          if (x >= (a+1.0)) {
               gcf(&gammcf,a,x,&gln);
               printf("%6.2f%13.6f%13.6f%13.6f%12.6f%13.6f\n",
                    a,x,(1.0-val),gammcf,gammln(a),gln);
          }
     }
     fclose(fp);
}
```

Error function:

```
/* Driver for routine ERF */

#include <stdio.h>
#include "nr.h"
#include "nrutil.h"

#define MAXSTR 80

main()
{
     char txt[MAXSTR];
     int i,nval,strncmp();
     float val,x;
     FILE *fp;

     if ((fp = fopen("fncval.dat","r")) == NULL)
          nrerror("Data file FNCVAL.DAT not found\n");
     fgets(txt,MAXSTR,fp);
     while(strncmp(txt,"Error Function",14)) {
          fgets(txt,MAXSTR,fp);
          if (feof(fp)) return;
     }
     fscanf(fp,"%d %*s",&nval);
     printf("\n%s\n",txt);
     printf("%4s %12s %12s\n","x","actual","erf(x)");
     for (i=1;i<=nval;i++) {
          fscanf(fp,"%f %f",&x,&val);
          printf("%6.2f %12.7f %12.7f\n",x,val,erf(x));
     }
     fclose(fp);
}
```

Complementary error function:

```
/* Driver for routine ERFC */

#include <stdio.h>
#include "nr.h"
#include "nrutil.h"

#define MAXSTR 80

main()
{
    char txt[MAXSTR];
    int i,nval,strncmp();
    float val,x;
    FILE *fp;

    if ((fp = fopen("fncval.dat","r")) == NULL)
        nrerror("Data file FNCVAL.DAT not found\n");
    fgets(txt,MAXSTR,fp);
    while(strncmp(txt,"Error Function",14)) {
        fgets(txt,MAXSTR,fp);
        if (feof(fp)) return;
    }
    fscanf(fp,"%d %*s",&nval);
    printf("\nComplementary error function\n");
    printf("%5s %12s %12s\n","x","actual","erfc(x)");
    for (i=1;i<=nval;i++) {
        fscanf(fp,"%f %f",&x,&val);
        val=1.0-val;
        printf("%6.2f %12.7f %12.7f\n",x,val,erfc(x));
    }
    fclose(fp);
}
```

Complementary error function from a Chebyshev fit to a guessed functional form:

```
/* Driver for routine ERFCC */

#include <stdio.h>
#include "nr.h"
#include "nrutil.h"

#define MAXSTR 80

main()
{
    char txt[MAXSTR];
    int i,nval,strncmp();
    float x,val;
    FILE *fp;

    if ((fp = fopen("fncval.dat","r")) == NULL)
        nrerror("Data file FNCVAL.DAT not found\n");
    fgets(txt,MAXSTR,fp);
    while(strncmp(txt,"Error Function",14))     {
        fgets(txt,MAXSTR,fp);
        if (feof(fp)) return;
```

```
    }
    fscanf(fp,"%d %*s",&nval);
    printf("\ncomplementary error function\n");
    printf("%5s %12s %13s\n","x","actual","erfcc(x)");
    for (i=1;i<=nval;i++) {
        fscanf(fp,"%f %f",&x,&val);
        val=1.0-val;
        printf("%6.2f %12.7f %12.7f\n",x,val,erfcc(x));
    }
    fclose(fp);
}
```

Incomplete Beta function:

```
/* Driver for routine BETAI */

#include <stdio.h>
#include "nr.h"
#include "nrutil.h"

#define MAXSTR 80

main()
{
    char txt[MAXSTR];
    int i,nval,strncmp();
    float a,b,x,val;
    FILE *fp;

    if ((fp = fopen("fncval.dat","r")) == NULL)
        nrerror("Data file FNCVAL.DAT not found\n");
    fgets(txt,MAXSTR,fp);
    while(strncmp(txt,"Incomplete Beta Function",24)) {
        fgets(txt,MAXSTR,fp);
        if (feof(fp)) return;
    }
    fscanf(fp,"%d %*s",&nval);
    printf("\n%s\n",txt);
    printf("%5s %10s %12s %14s %13s \n",
        "a","b","x","actual","betai(x)");
    for (i=1;i<=nval;i++) {
        fscanf(fp,"%f %f %f %f",&a,&b,&x,&val);
        printf("%6.2f %12.6f %12.6f %12.6f %12.6f\n",
            a,b,x,val,betai(a,b,x));
    }
    fclose(fp);
}
```

Bessel function J_0:

```
/* Driver for BESSJ0 */

#include <stdio.h>
#include "nr.h"
#include "nrutil.h"

#define MAXSTR 80

main()
```

```
{
    char txt[MAXSTR];
    int i,nval,strncmp();
    float val,x;
    FILE *fp;

    if ((fp = fopen("fncval.dat","r")) == NULL)
        nrerror("Data file FNCVAL.DAT not found\n");
    fgets(txt,MAXSTR,fp);
    while (strncmp(txt,"Bessel Function J0",18)) {
        fgets(txt,MAXSTR,fp);
        if (feof(fp)) return;
    }
    fscanf(fp,"%d %*s",&nval);
    printf("\n%s\n",txt);
    printf("%5s %12s %13s \n","x","actual","bessj0(x)");
    for (i=1;i<=nval;i++) {
        fscanf(fp,"%f %f",&x,&val);
        printf("%6.2f %12.7f %12.7f \n",x,val,bessj0(x));
    }
    fclose(fp);
}
```

Bessel function Y_0:

```
/* Driver for routine BESSY0 */

#include <stdio.h>
#include "nr.h"
#include "nrutil.h"

#define MAXSTR 80

main()
{
    char txt[MAXSTR];
    int i,nval,strncmp();
    float val,x;
    FILE *fp;

    if ((fp = fopen("fncval.dat","r")) == NULL)
        nrerror("Data file FNCVAL.DAT not found\n");
    fgets(txt,MAXSTR,fp);
    while(strncmp(txt,"Bessel Function Y0",18)) {
        fgets(txt,MAXSTR,fp);
        if (feof(fp)) return;
    }
    fscanf(fp,"%d %*s",&nval);
    printf("\n%s\n",txt);
    printf("%5s %12s %13s \n","x","actual","bessy0(x)");
    for (i=1;i<=nval;i++) {
        fscanf(fp,"%f %f",&x,&val);
        printf("%6.2f %12.7f %12.7f\n",x,val,bessy0(x));
    }
    fclose(fp);
}
```

Bessel function J_1:

```
/* Driver for routine BESSJ1 */

#include <stdio.h>
#include "nr.h"
#include "nrutil.h"

#define MAXSTR 80

main()
{
    char txt[MAXSTR];
    int i,nval,strncmp();
    float val,x;
    FILE *fp;

    if ((fp = fopen("fncval.dat","r")) == NULL)
        nrerror("Data file FNCVAL.DAT not found\n");
    fgets(txt,MAXSTR,fp);
    while(strncmp(txt,"Bessel Function J1",18)) {
        fgets(txt,MAXSTR,fp);
        if (feof(fp)) return;
    }
    fscanf(fp,"%d %*s",&nval);
    printf("\n%s\n",txt);
    printf("%5s %12s %13s \n","x","actual","bessj1(x)");
    for (i=1;i<=nval;i++) {
        fscanf(fp,"%f %f",&x,&val);
        printf("%6.2f %12.7f %12.7f\n",x,val,bessj1(x));
    }
    fclose(fp);
}
```

Bessel function Y_1:

```
/* Driver for routine BESSY1 */

#include <stdio.h>
#include "nr.h"
#include "nrutil.h"

#define MAXSTR 80

main()
{
    char txt[MAXSTR];
    int i,nval,strncmp();
    float val,x;
    FILE *fp;

    if ((fp = fopen("fncval.dat","r")) == NULL)
        nrerror("Data file FNCVAL.DAT not found\n");
    fgets(txt,MAXSTR,fp);
    while(strncmp(txt,"Bessel Function Y1",18)) {
        fgets(txt,MAXSTR,fp);
        if (feof(fp)) return;
    }
    fscanf(fp,"%d %*s",&nval);
    printf("\n%s\n",txt);
```

```
        printf("%5s %12s %13s \n","x","actual","bessy1(x)");
        for (i=1;i<=nval;i++) {
            fscanf(fp,"%f %f",&x,&val);
            printf("%6.2f %12.7f %12.7f\n",x,val,bessy1(x));
        }
        fclose(fp);
}
```

Bessel function Y_n for $n > 1$:

```
/* Driver for routine BESSY */

#include <stdio.h>
#include "nr.h"
#include "nrutil.h"

#define MAXSTR 80

main()
{
    char txt[MAXSTR];
    int i,nval,n,strncmp();
    float val,x;
    FILE *fp;

    if ((fp = fopen("fncval.dat","r")) == NULL)
        nrerror("Data file FNCVAL.DAT not found\n");
    fgets(txt,MAXSTR,fp);
    while(strncmp(txt,"Bessel Function Yn",18)) {
        fgets(txt,MAXSTR,fp);
        if (feof(fp)) return;
    }
    fscanf(fp,"%d %*s",&nval);
    printf("\n%s\n",txt);
    printf("%4s %7s %15s %20s \n","n","x","actual","bessy(n,x)");
    for (i=1;i<=nval;i++) {
        fscanf(fp,"%d %f %f",&n,&x,&val);
        printf("%4d %8.2f %18.6e %18.6e\n",n,x,val,bessy(n,x));
    }
    fclose(fp);
}
```

Bessel function J_n for $n > 1$:

```
/* Driver for routine BESSJ */

#include <stdio.h>
#include "nr.h"
#include "nrutil.h"

#define MAXSTR 80

main()
{
    char txt[MAXSTR];
    int i,nval,n,strncmp();
    float val,x;
    FILE *fp;
```

```
    if ((fp = fopen("fncval.dat","r")) == NULL)
        nrerror("Data file FNCVAL.DAT not found\n");
    fgets(txt,MAXSTR,fp);
    while (strncmp(txt,"Bessel Function Jn",18)) {
        fgets(txt,MAXSTR,fp);
        if (feof(fp)) return;
    }
    fscanf(fp,"%d %*s",&nval);
    printf("\n%s\n",txt);
    printf("%4s %7s %15s %20s \n","n","x","actual","bessj(n,x)");
    for (i=1;i<=nval;i++) {
        fscanf(fp,"%d %f %f",&n,&x,&val);
        printf("%4d %8.2f %18.6e %18.6e\n",n,x,val,bessj(n,x));
    }
    fclose(fp);
}
```

Bessel function I_0:

```
/* Driver for routine BESSI0 */

#include <stdio.h>
#include "nr.h"
#include "nrutil.h"

#define MAXSTR 80

main()
{
    char txt[MAXSTR];
    int i,nval,strncmp();
    float val,x;
    FILE *fp;

    if ((fp = fopen("fncval.dat","r")) == NULL)
        nrerror("Data file FNCVAL.DAT not found\n");
    fgets(txt,MAXSTR,fp);
    while(strncmp(txt,"Modified Bessel Function I0",27)) {
        fgets(txt,MAXSTR,fp);
        if (feof(fp)) return;
    }
    fscanf(fp,"%d %*s",&nval);
    printf("\n%s\n",txt);
    printf("%5s %12s %13s \n","x","actual","bessi0(x)");
    for (i=1;i<=nval;i++) {
        fscanf(fp,"%f %f",&x,&val);
        printf("%6.2f %12.7f %12.7f\n",x,val,bessi0(x));
    }
    fclose(fp);
}
```

Bessel function K_0:

```
/* Driver for routine BESSK0 */

#include <stdio.h>
#include "nr.h"
#include "nrutil.h"
```

```
#define MAXSTR 80

main()
{
    char txt[MAXSTR];
    int i,nval,strncmp();
    float val,x;
    FILE *fp;

    if ((fp = fopen("fncval.dat","r")) == NULL)
        nrerror("Data file FNCVAL.DAT not found\n");
    fgets(txt,MAXSTR,fp);
    while(strncmp(txt,"Modified Bessel Function K0",27)) {
        fgets(txt,MAXSTR,fp);
        if (feof(fp)) return;
    }
    fscanf(fp,"%d %*s",&nval);
    printf("\n%s\n",txt);
    printf("%5s %13s %18s \n","x","actual","bessk0(x)");
    for (i=1;i<=nval;i++) {
        fscanf(fp,"%f %f",&x,&val);
        printf("%6.2f %16.7e %16.7e\n",x,val,bessk0(x));
    }
    fclose(fp);
}
```

Bessel function I_1:

```
/* Driver for routine BESSI1 */

#include <stdio.h>
#include "nr.h"
#include "nrutil.h"

#define MAXSTR 80

main()
{
    char txt[MAXSTR];
    int i,nval,strncmp();
    float val,x;
    FILE *fp;

    if ((fp = fopen("fncval.dat","r")) == NULL)
        nrerror("Data file FNCVAL.DAT not found\n");
    fgets(txt,MAXSTR,fp);
    while(strncmp(txt,"Modified Bessel Function I1",27)) {
        fgets(txt,MAXSTR,fp);
        if (feof(fp)) return;
    }
    fscanf(fp,"%d %*s",&nval);
    printf("\n%s\n",txt);
    printf("%5s %12s %13s \n","x","actual","bessi1(x)");
    for (i=1;i<=nval;i++) {
        fscanf(fp,"%f %f",&x,&val);
        printf("%6.2f %12.7f %12.7f\n",x,val,bessi1(x));
    }
    fclose(fp);
```

```
}
```

Bessel function K_1:

```
/* Driver for routine BESSK1 */

#include <stdio.h>
#include "nr.h"
#include "nrutil.h"

#define MAXSTR 80

main()
{
    char txt[MAXSTR];
    int i,nval,strncmp();
    float val,x;
    FILE *fp;

    if ((fp = fopen("fncval.dat","r")) == NULL)
        nrerror("Data file FNCVAL.DAT not found\n");
    fgets(txt,MAXSTR,fp);
    while(strncmp(txt,"Modified Bessel Function K1",27)) {
        fgets(txt,MAXSTR,fp);
        if (feof(fp)) return;
    }
    fscanf(fp,"%d %*s",&nval);
    printf("\n%s\n",txt);
    printf("%5s %13s %17s \n","x","actual","bessk1(x)");
    for (i=1;i<=nval;i++) {
        fscanf(fp,"%f %f",&x,&val);
        printf("%6.2f %16.7e %16.7e\n",x,val,bessk1(x));
    }
    fclose(fp);
}
```

Bessel function K_n for $n > 1$:

```
/* Driver for routine BESSK */

#include <stdio.h>
#include "nr.h"
#include "nrutil.h"

#define MAXSTR 80

main()
{
    char txt[MAXSTR];
    int i,nval,n,strncmp();
    float val,x;
    FILE *fp;

    if ((fp = fopen("fncval.dat","r")) == NULL)
        nrerror("Data file FNCVAL.DAT not found\n");
    fgets(txt,MAXSTR,fp);
    while(strncmp(txt,"Modified Bessel Function Kn",27)) {
        fgets(txt,MAXSTR,fp);
        if (feof(fp)) return;
```

```
    }
    fscanf(fp,"%d %*s",&nval);
    printf("\n%s\n",txt);
    printf("%4s %7s %14s %19s\n","n","x","actual","bessk(n,x)");
    for (i=1;i<=nval;i++) {
        fscanf(fp,"%d %f %f",&n,&x,&val);
        printf("%4d %8.2f %18.7e %16.7e \n",n,x,val,bessk(n,x));
    }
    fclose(fp);
}
```

Bessel function I_n for $n > 1$:

```
/* Driver for routine BESSI */

#include <stdio.h>
#include "nr.h"
#include "nrutil.h"

#define MAXSTR 80

main()
{
    char txt[MAXSTR];
    int i,nval,n,strncmp();
    float val,x;
    FILE *fp;

    if ((fp = fopen("fncval.dat","r")) == NULL)
        nrerror("Data file FNCVAL.DAT not found\n");
    fgets(txt,MAXSTR,fp);
    while(strncmp(txt,"Modified Bessel Function In",27)) {
        fgets(txt,MAXSTR,fp);
        if (feof(fp)) return;
    }
    fscanf(fp,"%d %*s",&nval);
    printf("\n%s\n",txt);
    printf("%4s %7s %15s %20s\n","n","x","actual","bessi(n,x)");
    for (i=1;i<=nval;i++) {
        fscanf(fp,"%d %f %f",&n,&x,&val);
        printf("%4d %8.2f %18.7e %18.7e\n",n,x,val,bessi(n,x));
    }
    fclose(fp);
}
```

Legendre polynomials:

```
/* Driver for routine PLGNDR */

#include <stdio.h>
#include <math.h>
#include "nr.h"
#include "nrutil.h"

#define MAXSTR 80

main()
{
    char txt[MAXSTR];
```

```
        int i,j,m,n,nval,strncmp();
        float fac,val,x;
        FILE *fp;

        if ((fp = fopen("fncval.dat","r")) == NULL)
            nrerror("Data file FNCVAL.DAT not found\n");
        fgets(txt,MAXSTR,fp);
        while(strncmp(txt,"Legendre Polynomials",20)) {
            fgets(txt,MAXSTR,fp);
            if (feof(fp)) return;
        }
        fscanf(fp,"%d %*s",&nval);
        printf("\n%s\n",txt);
        printf("%4s %4s %10s %17s %24s\n","n",
            "m","x","actual","plgndr(n,m,x)");
        for (i=1;i<=nval;i++) {
            fscanf(fp,"%d %d %f %f",&n,&m,&x,&val);
            fac=1.0;
            if (m > 0)
                for (j=n-m+1;j<=n+m;j++) fac *= j;
            fac *= 2.0/(2.0*n+1.0);
            val *= sqrt(fac);
            printf("%4d %4d %13.6f %19.6e %19.6e\n",
                n,m,x,val,plgndr(n,m,x));
        }
        fclose(fp);
}
```

There are three programs that operate in a slightly different fashion. Sample programs
xel2.c and xcel.c for routines el2 and cel, which calculate elliptic integrals, do
not refer to tables at all. Instead, they make twenty random choices of argument and
compare the output of the function evaluation routine for these arguments to the result of
actually performing the integration that defines them. The routine qsimp from Chapter
4 of *Numerical Recipes* is used for the integration.

```
/* Driver for routine EL2 */

#include <stdio.h>
#include <math.h>
#include "nr.h"

static float a,b,akc;

float func(phi)
float phi;
{
    float tn,tsq;

    tn=tan(phi);
    tsq=tn*tn;
    return (a+b*tsq)/sqrt((1.0+tsq)*(1.0+akc*akc*tsq));
}

main()
{
    float astop,s,x,ago=0.0;
    float func();
```

```
    int i,idum=(-55);


    printf("general elliptic integral of second kind\n");
    printf("%7s %10s %10s %10s %11s %12s\n",
        "x","kc","a","b","el2","integral");
    for (i=1;i<=20;i++) {
        akc=5.0*ran3(&idum);
        a=10.0*ran3(&idum);
        b=10.0*ran3(&idum);
        x=10.0*ran3(&idum);
        astop=atan(x);
        s=qsimp(func,ago,astop);
        printf("%10.6f %10.6f %10.6f %10.6f %10.6f %10.6f\n",
            x,akc,a,b,el2(x,akc,a,b),s);
    }
}

/* Driver for routine CEL */

#include <stdio.h>
#include <math.h>
#include "nr.h"

#define PIO2 1.5707963

static float a,b,p,akc;

float func(phi)
float phi;
{
    float cs,csq,ssq;

    cs=cos(phi);
    csq=cs*cs;
    ssq=1.0-csq;
    return (a*csq+b*ssq)/(csq+p*ssq)/sqrt(csq+akc*akc*ssq);
}

main()
{
    float s,ago=0.0,astop=PIO2;
    float func();
    int i,idum=(-55);

    printf("Incomplete elliptic integral\n");
    printf("%7s %10s %10s %10s %11s %12s\n",
        "kc","p","a","b","cel","integral");
    for (i=1;i<=20;i++) {
        akc=0.1+ran3(&idum);
        a=10.0*ran3(&idum);
        b=10.0*ran3(&idum);
        p=0.1+ran3(&idum);
        s=qsimp(func,ago,astop);
        printf("%10.6f %10.6f %10.6f %10.6f %10.6f %10.6f\n",
            akc,p,a,b,cel(akc,p,a,b),s);
    }
```

}

Routine sncndn returns Jacobian elliptic functions. The file fncval.dat contains information only about function sn. However, the values of cn and dn satisfy the relationships

$$\text{sn}^2 + \text{cn}^2 = 1, \qquad k^2 \text{sn}^2 + \text{dn}^2 = 1.$$

The program xsncndn.c works exactly as the others in terms of testing sn, but for verifying cn and dn it lists the values of the left sides of the two equations above. Each of them should have the value 1.0 for all choices of arguments.

```c
/* Driver for routine SNCNDN */

#include <stdio.h>
#include "nr.h"
#include "nrutil.h"

#define MAXSTR 80

main()
{
    char txt[MAXSTR];
    int i,nval,strncmp();
    float em,emmc,uu,val,sn,cn,dn;
    FILE *fp;

    if ((fp = fopen("fncval.dat","r")) == NULL)
        nrerror("Data file FNCVAL.DAT not found\n");
    fgets(txt,MAXSTR,fp);
    while(strncmp(txt,"Jacobian Elliptic Function",26)) {
        fgets(txt,MAXSTR,fp);
        if (feof(fp)) return;
    }
    fscanf(fp,"%d %*s",&nval);
    printf("\n%s\n",txt);
    printf("%4s %8s %16s %13s %15s %18s\n","mc","u","actual",
        "sn","sn^2+cn^2","(mc)*(sn^2)+dn^2");
    for (i=1;i<=nval;i++) {
        fscanf(fp,"%f %f %f",&em,&uu,&val);
        emmc=1.0-em;
        sncndn(uu,emmc,&sn,&cn,&dn);
        printf("%5.2f %8.2f %15.5f %15.5f %12.5f %14.5f\n",
            emmc,uu,val,sn,(sn*sn+cn*cn),(em*sn*sn+dn*dn));
    }
    fclose(fp);
}
```

Appendix

File fncval.dat:

```
Values of Special Functions in format x,F(x) or x,y,F(x,y)
Gamma Function
17 Values
   1.0    1.000000
   1.2    0.918169
   1.4    0.887264
   1.6    0.893515
```

```
 1.8     0.931384
 2.0     1.000000
 0.2     4.590845
 0.4     2.218160
 0.6     1.489192
 0.8     1.164230
-0.2     5.2005665E01
-0.4     4.617091E01
-0.6     4.0128959E01
-0.8     3.4231564E01
10.0     3.6288000E05
20.0     1.2164510E17
30.0     8.8417620E30
N-factorial
18 Values
1        1
2        2
3        6
4        24
5        120
6        720
7        5040
8        40320
9        362880
10       3628800
11       39916800
12       479001600
13       6227020800
14       87178291200
15       1.3076755E12
20       2.4329042E18
25       1.5511222E25
30       2.6525281E32
Binomial Coefficients
20 Values
1        0        1
6        1        6
6        3        20
6        5        6
15       1        15
15       3        455
15       5        3003
15       7        6435
15       9        5005
15       11       1365
15       13       105
25       1        25
25       3        2300
25       5        53130
25       7        480700
25       9        2042975
25       11       4457400
25       13       5200300
25       15       3268760
25       17       1081575
Beta Function
15 Values
```

1.0	1.0	1.000000
0.2	1.0	5.000000
1.0	0.2	5.000000
0.4	1.0	2.500000
1.0	0.4	2.500000
0.6	1.0	1.666667
0.8	1.0	1.250000
6.0	6.0	3.607504E-04
6.0	5.0	7.936508E-04
6.0	4.0	1.984127E-03
6.0	3.0	5.952381E-03
6.0	2.0	0.238095E-01
7.0	7.0	8.325008E-05
5.0	5.0	1.587302E-03
4.0	4.0	7.142857E-03
3.0	3.0	0.333333E-01
2.0	2.0	1.666667E-01

Incomplete Gamma Function

20 Values

0.1	3.1622777E-02	0.7420263
0.1	3.1622777E-01	0.9119753
0.1	1.5811388	0.9898955
0.5	7.0710678E-02	0.2931279
0.5	7.0710678E-01	0.7656418
0.5	3.5355339	0.9921661
1.0	0.1000000	0.0951626
1.0	1.0000000	0.6321206
1.0	5.0000000	0.9932621
1.1	1.0488088E-01	0.0757471
1.1	1.0488088	0.6076457
1.1	5.2440442	0.9933425
2.0	1.4142136E-01	0.0091054
2.0	1.4142136	0.4130643
2.0	7.0710678	0.9931450
6.0	2.4494897	0.0387318
6.0	12.247449	0.9825937
11.0	16.583124	0.9404267
26.0	25.495098	0.4863866
41.0	44.821870	0.7359709

Error Function

20 Values

0.0	0.000000
0.1	0.1124629
0.2	0.2227026
0.3	0.3286268
0.4	0.4283924
0.5	0.5204999
0.6	0.6038561
0.7	0.6778012
0.8	0.7421010
0.9	0.7969082
1.0	0.8427008
1.1	0.8802051
1.2	0.9103140
1.3	0.9340079
1.4	0.9522851
1.5	0.9661051

```
1.6     0.9763484
1.7     0.9837905
1.8     0.9890905
1.9     0.9927904
Incomplete Beta Function
20 Values
0.5     0.5     0.01    0.0637686
0.5     0.5     0.10    0.2048328
0.5     0.5     1.00    1.0000000
1.0     0.5     0.01    0.0050126
1.0     0.5     0.10    0.0513167
1.0     0.5     1.00    1.0000000
1.0     1.0     0.5     0.5000000
5.0     5.0     0.5     0.5000000
10.0    0.5     0.9     0.1516409
10.0    5.0     0.5     0.0897827
10.0    5.0     1.0     1.0000000
10.0    10.0    0.5     0.5000000
20.0    5.0     0.8     0.4598773
20.0    10.0    0.6     0.2146816
20.0    10.0    0.8     0.9507365
20.0    20.0    0.5     0.5000000
20.0    20.0    0.6     0.8979414
30.0    10.0    0.7     0.2241297
30.0    10.0    0.8     0.7586405
40.0    20.0    0.7     0.7001783
Bessel Function J0
20 Values
−5.0    −0.1775968
−4.0    −0.3971498
−3.0    −0.2600520
−2.0     0.2238908
−1.0     0.7651976
 0.0     1.0000000
 1.0     0.7651977
 2.0     0.2238908
 3.0    −0.2600520
 4.0    −0.3971498
 5.0    −0.1775968
 6.0     0.1506453
 7.0     0.3000793
 8.0     0.1716508
 9.0    −0.0903336
10.0    −0.2459358
11.0    −0.1711903
12.0     0.0476893
13.0     0.2069261
14.0     0.1710735
15.0    −0.0142245
Bessel Function Y0
15 Values
 0.1    −1.5342387
 1.0     0.0882570
 2.0     0.51037567
 3.0     0.37685001
 4.0    −0.0169407
 5.0    −0.3085176
```

```
 6.0   -0.2881947
 7.0   -0.0259497
 8.0    0.2235215
 9.0    0.2499367
10.0    0.0556712
11.0   -0.1688473
12.0   -0.2252373
13.0   -0.0782079
14.0    0.1271926
15.0    0.2054743
Bessel Function J1
20 Values
-5.0    0.3275791
-4.0    0.0660433
-3.0   -0.3390590
-2.0   -0.5767248
-1.0   -0.4400506
 0.0    0.0000000
 1.0    0.4400506
 2.0    0.5767248
 3.0    0.3390590
 4.0   -0.0660433
 5.0   -0.3275791
 6.0   -0.2766839
 7.0   -0.0046828
 8.0    0.2346364
 9.0    0.2453118
10.0    0.0434728
11.0   -0.1767853
12.0   -0.2234471
13.0   -0.0703181
14.0    0.1333752
15.0    0.2051040
Bessel Function Y1
15 Values
 0.1   -6.4589511
 1.0   -0.7812128
 2.0   -0.1070324
 3.0    0.3246744
 4.0    0.3979257
 5.0    0.1478631
 6.0   -0.1750103
 7.0   -0.3026672
 8.0   -0.1580605
 9.0    0.1043146
10.0    0.2490154
11.0    0.1637055
12.0   -0.0570992
13.0   -0.2100814
14.0   -0.1666448
15.0    0.0210736
Bessel Function Jn, n>=2
20 Values
2       1.0     1.149034849E-01
2       2.0     3.528340286E-01
2       5.0     4.656511628E-02
2      10.0     2.546303137E-01
```

```
2       50.0     -5.971280079E-02
5        1.0      2.497577302E-04
5        2.0      7.039629756E-03
5        5.0      2.611405461E-01
5       10.0     -2.340615282E-01
5       50.0     -8.140024770E-02
10       1.0      2.630615124E-10
10       2.0      2.515386283E-07
10       5.0      1.467802647E-03
10      10.0      2.074861066E-01
10      50.0     -1.138478491E-01
20       1.0      3.873503009E-25
20       2.0      3.918972805E-19
20       5.0      2.770330052E-11
20      10.0      1.151336925E-05
20      50.0     -1.167043528E-01
```

Bessel Function Yn, n>=2
20 Values

```
2        1.0     -1.650682607
2        2.0     -6.174081042E-01
2        5.0      3.676628826E-01
2       10.0     -5.868082460E-03
2       50.0      9.579316873E-02
5        1.0     -2.604058666E02
5        2.0     -9.935989128
5        5.0     -4.536948225E-01
5       10.0      1.354030477E-01
5       50.0     -7.854841391E-02
10       1.0     -1.216180143E08
10       2.0     -1.291845422E05
10       5.0     -2.512911010E01
10      10.0     -3.598141522E-01
10      50.0      5.723897182E-03
20       1.0     -4.113970315E22
20       2.0     -4.081651389E16
20       5.0     -5.933965297E08
20      10.0     -1.597483848E03
20      50.0      1.644263395E-02
```

Modified Bessel Function I0
20 Values

```
0.0      1.0000000
0.2      1.0100250
0.4      1.0404018
0.6      1.0920453
0.8      1.1665149
1.0      1.2660658
1.2      1.3937256
1.4      1.5533951
1.6      1.7499807
1.8      1.9895593
2.0      2.2795852
2.5      3.2898391
3.0      4.8807925
3.5      7.3782035
4.0     11.301922
4.5     17.481172
5.0     27.239871
```

```
6.0     67.234406
8.0     427.56411
10.0    2815.7167
Modified Bessel Function K0
20 Values
0.1     2.4270690
0.2     1.7527038
0.4     1.1145291
0.6     0.77752208
0.8     0.56534710
1.0     0.42102445
1.2     0.31850821
1.4     0.24365506
1.6     0.18795475
1.8     0.14593140
2.0     0.11389387
2.5     6.2347553E-02
3.0     3.4739500E-02
3.5     1.9598897E-02
4.0     1.1159676E-02
4.5     6.3998572E-03
5.0     3.6910983E-03
6.0     1.2439943E-03
8.0     1.4647071E-04
10.0    1.7780062E-05
Modified Bessel Function I1
20 Values
0.0     0.00000000
0.2     0.10050083
0.4     0.20402675
0.6     0.31370403
0.8     0.43286480
1.0     0.56515912
1.2     0.71467794
1.4     0.88609197
1.6     1.0848107
1.8     1.3171674
2.0     1.5906369
2.5     2.5167163
3.0     3.9533700
3.5     6.2058350
4.0     9.7594652
4.5     15.389221
5.0     24.335643
6.0     61.341937
8.0     399.87313
10.0    2670.9883
Modified Bessel Function K1
20 Values
0.1     9.8538451
0.2     4.7759725
0.4     2.1843544
0.6     1.3028349
0.8     0.86178163
1.0     0.60190724
1.2     0.43459241
1.4     0.32083589
```

```
1.6    0.24063392
1.8    0.18262309
2.0    0.13986588
2.5    7.3890816E-02
3.0    4.0156431E-02
3.5    2.2239393E-02
4.0    1.2483499E-02
4.5    7.0780949E-03
5.0    4.0446134E-03
6.0    1.3439197E-03
8.0    1.5536921E-04
10.0   1.8648773E-05
Modified Bessel Function Kn, n>=2
28 Values
2      0.2      49.512430
2      1.0      1.6248389
2      2.0      2.5375975E-01
2      2.5      1.2146021E-01
2      3.0      6.1510459E-02
2      5.0      5.3089437E-03
2      10.0     2.1509817E-05
2      20.0     6.3295437E-10
3      1.0      7.101262825
3      2.0      6.473853909E-01
3      5.0      8.291768415E-03
3      10.0     2.725270026E-05
3      50.0     3.72793677E-23
5      1.0      3.609605896E02
5      2.0      9.431049101
5      5.0      3.270627371E-02
5      10.0     5.754184999E-05
5      50.0     4.36718224E-23
10     1.0      1.807132899E08
10     2.0      1.624824040E05
10     5.0      9.758562829
10     10.0     1.614255300E-03
10     50.0     9.15098819E-23
20     1.0      6.294369369E22
20     2.0      5.770856853E16
20     5.0      4.827000521E08
20     10.0     1.787442782E02
20     50.0     1.70614838E-21
Modified Bessel Function In, n>=2
28 Values
2      0.2      5.0166876E-03
2      1.0      1.3574767E-01
2      2.0      6.8894844E-01
2      2.5      1.2764661
2      3.0      2.2452125
2      5.0      17.505615
2      10.0     2281.5189
2      20.0     3.9312785E07
3      1.0      2.216842492E-02
3      2.0      2.127399592E-01
3      5.0      1.033115017E01
3      10.0     1.758380717E01
3      50.0     2.67776414E20
```

```
5      1.0      2.714631560E-04
5      2.0      9.825679323E-03
5      5.0      2.157974547
5      10.0     7.771882864E02
5      50.0     2.27854831E20
10     1.0      2.752948040E-10
10     2.0      3.016963879E-07
10     5.0      4.580044419E-03
10     10.0     2.189170616E01
10     50.0     1.07159716E20
20     1.0      3.966835986E-25
20     2.0      4.310560576E-19
20     5.0      5.024239358E-11
20     10.0     1.250799736E-04
20     50.0     5.44200840E18
```

Legendre Polynomials
19 Values
```
1      0      1.0          1.224745
10     0      1.0          3.240370
20     0      1.0          4.527693
1      0      0.7071067    0.866025
10     0      0.7071067    0.373006
20     0      0.7071067   -0.874140
1      0      0.0          0.000000
10     0      0.0         -0.797435
20     0      0.0          0.797766
2      2      0.7071067    0.484123
10     2      0.7071067   -0.204789
20     2      0.7071067    0.910208
2      2      0.0          0.968246
10     2      0.0          0.804785
20     2      0.0         -0.799672
10     10     0.7071067    0.042505
20     10     0.7071067   -0.707252
10     10     0.0          1.360172
20     10     0.0         -0.853705
```

Jacobian Elliptic Function
20 Values
```
0.0    0.1     0.099833
0.0    0.2     0.19867
0.0    0.5     0.47943
0.0    1.0     0.84147
0.0    2.0     0.90930
0.5    0.1     0.099751
0.5    0.2     0.19802
0.5    0.5     0.47075
0.5    1.0     0.80300
0.5    2.0     0.99466
1.0    0.1     0.099668
1.0    0.2     0.19738
1.0    0.5     0.46212
1.0    1.0     0.76159
1.0    2.0     0.96403
1.0    4.0     0.99933
1.0   -0.2    -0.19738
1.0   -0.5    -0.46212
1.0   -1.0    -0.76159
```

```
1.0    -2.0    -0.96403
```

Chapter 7: Random Numbers

Chapter 7 of Numerical Recipes deals with the generation of random numbers drawn from various distributions. The first four procedures produce uniform deviates with a range of 0.0 to 1.0. ran0 *is a procedure for improving the randomness of a system-supplied random number generator by shuffling the output.* ran1 *is a portable random number generator based on three linear congruential generators and a shuffler.* ran2 *contains a single congruential generator and a shuffler, and has the advantage of being somewhat faster, if less plentiful in possible output values.* ran3 *is another portable generator, based on a subtractive rather than a congruential method.*

The transformation method is used to generate some non-uniform distributions. Resulting from this are routines expdev, *which gives exponentially distributed deviates, and* gasdev *for Gaussian deviates. The rejection method of producing non-uniform deviates is also discussed, and is used in* gamdev *(deviate with a gamma function distribution),* poidev *(deviate with a Poisson distribution), and* bnldev *(deviate with a binomial distribution). For generating random sequences of zeros and ones, there are two procedures,* irbit1 *and* irbit2, *both based on a 32-bit seed* iseed, *but each using a different recurrence to proceed from step to step. The national Data Encryption Standard (DES) is discussed as the basis for a random number generator which we call* ran4. *DES itself is carried out by routines* des, ks, *and* cyfun.

★   ★   ★   ★

The first four sample programs in this chapter are really all the same, except that each calls a different random number generator (ran0 to ran3, respectively). They first draw four consecutive random numbers $X_1, \ldots, X_4$ from the generator in question. Then they treat the numbers as coordinates of a point. For example, they take (X_1, X_2) as a point in two dimensions, (X_1, X_2, X_3) as a point in three dimensions, etc. These points are inside boxes of unit dimension in their respective n-space. They may, however, be either inside or outside of the unit sphere in that space. For $n = 2, 3, 4$ we seek the probability that a point is inside the unit n-sphere. This number is easily calculated theoretically. For $n = 2$ it is $\pi/4$, for $n = 3$ it is $\pi/6$, and for $n = 4$ it is $\pi^2/32$. If the random number generator is not faulty, the points will fall within the unit n-sphere this fraction of the time, and the result should become increasingly accurate as the number of points increases. In these examples we have taken out a factor of 2^n for convenience, and used the random number generators as a statistical means of determining the value of π, $4\pi/3$, and $\pi^2/2$.

```
/* Driver for routine RAN0 */

#include <stdio.h>
#include <math.h>
#include "nr.h"

#define PI 3.1415926

int twotoj(j)
int j;
{
    return j == 0 ? 1 : 2*twotoj(j-1);
}

float fnc(x1,x2,x3,x4)
float x1,x2,x3,x4;
{
    return (float) sqrt(x1*x1+x2*x2+x3*x3+x4*x4);
}

main()
{
    int i,idum=(-1),j,k,jpower;
    float x1,x2,x3,x4;
    float iy[4],yprob[4];

    /* Calculates PI statistically using volume of unit n-sphere */
    for (i=1;i<=3;i++) iy[i]=0;
    printf("\nvolume of unit n-sphere, n = 2,3,4\n");
    printf("# points      PI         (4/3)*PI     (1/2)*PI^2\n\n");
    for (j=1;j<=14;j++) {
        for (k=twotoj(j-1);k<=twotoj(j);k++) {
            x1=ran0(&idum);
            x2=ran0(&idum);
            x3=ran0(&idum);
            x4=ran0(&idum);
            if (fnc(x1,x2,0.0,0.0) < 1.0) ++iy[1];
            if (fnc(x1,x2,x3,0.0) < 1.0) ++iy[2];
            if (fnc(x1,x2,x3,x4) < 1.0) ++iy[3];
        }
        jpower=twotoj(j);
        yprob[1]=4.0*iy[1]/jpower;
        yprob[2]=8.0*iy[2]/jpower;
        yprob[3]=16.0*iy[3]/jpower;
        printf("%6d %12.6f %12.6f %12.6f\n",
            jpower,yprob[1],yprob[2],yprob[3]);
    }
    printf("\nactual %12.6f %12.6f %12.6f\n",
        PI,4.0*PI/3.0,0.5*PI*PI);
}

/* Driver for routine RAN1 */

#include <stdio.h>
#include <math.h>
#include "nr.h"
```

```
#define PI 3.1415926

int twotoj(j)
int j;
{
    return j == 0 ? 1 : 2*twotoj(j-1);
}

float fnc(x1,x2,x3,x4)
float x1,x2,x3,x4;
{
    return (float) sqrt(x1*x1+x2*x2+x3*x3+x4*x4);
}

main()
{
    int i,idum=(-1),j,k,jpower;
    float x1,x2,x3,x4;
    float iy[4],yprob[4];

    /* Calculates PI statistically using volume of unit n-sphere */
    for (i=1;i<=3;i++) iy[i]=0;
    printf("\nvolume of unit n-sphere, n = 2,3,4\n");
    printf("# points      PI       (4/3)*PI    (1/2)*PI^2\n\n");
    for (j=1;j<=14;j++) {
        for (k=twotoj(j-1);k<=twotoj(j);k++) {
            x1=ran1(&idum);
            x2=ran1(&idum);
            x3=ran1(&idum);
            x4=ran1(&idum);
            if (fnc(x1,x2,0.0,0.0) < 1.0) ++iy[1];
            if (fnc(x1,x2,x3,0.0) < 1.0) ++iy[2];
            if (fnc(x1,x2,x3,x4) < 1.0) ++iy[3];
        }
        jpower=twotoj(j);
        yprob[1]=4.0*iy[1]/jpower;
        yprob[2]=8.0*iy[2]/jpower;
        yprob[3]=16.0*iy[3]/jpower;
        printf("%6d %12.6f %12.6f %12.6f\n",
            jpower,yprob[1],yprob[2],yprob[3]);
    }
    printf("\nactual %12.6f %12.6f %12.6f\n",
        PI,4.0*PI/3.0,0.5*PI*PI);
}
/* Driver for routine RAN2 */

#include <stdio.h>
#include <math.h>
#include "nr.h"

#define PI 3.1415926

float fnc(x1,x2,x3,x4)
float x1,x2,x3,x4;
{
    return (float) sqrt(x1*x1+x2*x2+x3*x3+x4*x4);
```

```
}

int twotoj(j)
int j;
{
    return j == 0 ? 1 : 2*twotoj(j-1);
}

main()
{
    long idum=(-1);
    int i,j,k,jpower,iy[4];
    float x1,x2,x3,x4,yprob[4];

    /* Calculates pi statistically using volume of unit n-sphere */
    for (i=1;i<=3;i++) iy[i]=0;
    printf("\nvolume of unit n-sphere, n=2,3,4\n");
    printf("# points      pi        (4/3)*pi   (1/2)*pi^2 \n\n");
    for (j=1;j<=14;j++) {
        for (k=twotoj(j-1);k<=twotoj(j);k++) {
            x1=ran2(&idum);
            x2=ran2(&idum);
            x3=ran2(&idum);
            x4=ran2(&idum);
            if (fnc(x1,x2,0.0,0.0) < 1.0) ++iy[1];
            if (fnc(x1,x2,x3,0.0) < 1.0) ++iy[2];
            if (fnc(x1,x2,x3,x4) < 1.0) ++iy[3];
        }
        jpower=twotoj(j);
        yprob[1]=4.0*iy[1]/jpower;
        yprob[2]=8.0*iy[2]/jpower;
        yprob[3]=16.0*iy[3]/jpower;
        printf("%6d %12.6f %12.6f %12.6f\n",
            jpower,yprob[1],yprob[2],yprob[3]);
    }
    printf("\nactual %12.6f %12.6f %12.6f\n",
        PI,4.0*PI/3.0,0.5*PI*PI);
}

/* Driver for routine RAN3 */

#include <stdio.h>
#include <math.h>
#include "nr.h"

#define PI 3.1415926

float fnc(x1,x2,x3,x4)
float x1,x2,x3,x4;
{
    return (float) sqrt(x1*x1+x2*x2+x3*x3+x4*x4);
}

int twotoj(j)
int j;
{
    return j == 0 ? 1 : 2*twotoj(j-1);
```

```
}
main()
{
    int i,j,k,idum=(-1),jpower;
    static int iy[4]={0,0,0,0};
    float x1,x2,x3,x4,yprob[4];

    /* Calculates pi statistically using volume of unit n-sphere */
    printf("\nvolume of unit n-sphere, n=2,3,4\n");
    printf(" # points      pi        (4/3)*pi    (1/2)*pi^2 \n\n");
    for (j=1;j<=14;j++) {
        for (k=twotoj(j-1);k<=twotoj(j);k++) {
            x1=ran3(&idum);
            x2=ran3(&idum);
            x3=ran3(&idum);
            x4=ran3(&idum);
            if (fnc(x1,x2,0.0,0.0) < 1.0) ++iy[1];
            if (fnc(x1,x2,x3,0.0) < 1.0) ++iy[2];
            if (fnc(x1,x2,x3,x4) < 1.0) ++iy[3];
        }
        jpower=twotoj(j);
        for (i=1;i<=3;i++)
            yprob[i]=(float) twotoj(i+1)*iy[i]/jpower;
        printf("%6d %12.6f %12.6f %12.6f\n",jpower,
            yprob[1],yprob[2],yprob[3]);
    }
    printf("\nactual %12.6f %12.6f %12.6f\n",
        PI,(4.0*PI/3.0),(0.5*PI*PI));
}
```

Routine `expdev` generates random numbers drawn from an exponential deviate. Sample program `xexpdev.c` makes two thousand calls to `expdev` and bins the results into 21 bins, the contents of which are tallied in array `x[i]`. Then the sum `total` of all bins is taken, since some of the numbers will be too large to have fallen in any of the bins. The `x[i]` are scaled to `total`, and then compared to a similarly normalized exponential which is called `expect`.

```
/* Driver for routine EXPDEV */

#include <stdio.h>
#include <math.h>
#include "nr.h"

#define NPTS 2000
#define EE 2.718281828

main()
{
    int i,idum,j,total,x[21];
    float expect,xx,y,trig[21];

    for (i=0;i<=20;i++) {
        trig[i]=i/20.0;
        x[i]=0;
    }
    idum=(-1);
```

```
    for (i=1;i<=NPTS;i++) {
        y=expdev(&idum);
        for (j=1;j<=20;j++)
            if ((y < trig[j]) && (y > trig[j-1])) ++x[j];
    }
    total=0;
    for (i=1;i<=20;i++) total += x[i];
    printf("\nexponential distribution with %7d points\n",NPTS);
    printf("    interval        observed        expected\n\n");
    for (i=1;i<=20;i++) {
        xx=(float) x[i]/total;
        expect=exp(-(trig[i-1]+trig[i])/2.0);
        expect *= (0.05*EE/(EE-1));
        printf("%6.2f %6.2f %12.6f %12.6f \n",
            trig[i-1],trig[i],xx,expect);
    }
}
```

gasdev generates random numbers from a Gaussian deviate. Example xgasdev.c takes two thousand of these and puts them into 21 bins. For the purpose of binning, the center of the Gaussian is shifted over by nover2=10 bins, to put it in the middle bin. The remainder of the program simply plots the contents of the bins, to illustrate that they have the characteristic Gaussian bell-shape. This allows a quick, though superficial, check of the integrity of the routine.

```
/* Driver for routine GASDEV */

#include <stdio.h>
#include "nr.h"

#define N 20
#define NOVER2 (N/2)
#define NPTS 2000
#define ISCAL 400
#define LLEN 50

main()
{
    char words[LLEN+1];
    int i,idum=(-13),j,k,klim,dist[N+1];
    float dd;

    for (j=0;j<=N;j++) dist[j]=0;
    for (i=1;i<=NPTS;i++) {
        j=(int) (0.5+0.25*N*gasdev(&idum));
        if ((j >= -NOVER2) && (j <= NOVER2)) ++dist[j+NOVER2];
    }
    printf("Normally distributed deviate of %6d points\n",NPTS);
    printf ("%5s %10s %9s\n","x","p(x)","graph:");
    for (j=0;j<=N;j++) {
        dd=(float) dist[j]/NPTS;
        for (k=1;k<=LLEN;k++) words[k]=' ';
        klim=ISCAL*dd;
        if (klim > LLEN)   klim=LLEN;
        for (k=1;k<=klim;k++) words[k]='*';
        printf("%8.4f %8.4f  ",j/(0.25*N),dd);
        for (k=1;k<=LLEN;k++) printf("%c",words[k]);
```

```
        printf("\n");
    }
}
```

The next three sample programs, xgamdev.c, xpoidev.c, and xbnldev.c, are identical to the previous one, but each drives a different random number generator and produces a different graph. xgamdev.c drives gamdev and displays a gamma distribution of order ia specified by the user. xpoidev.c drives poidev and produces a Poisson distribution with mean xm specified by the user. xbnldev.c drives bnldev and produces a binomial distribution, also with specified xm.

```
/* Driver for routine GAMDEV */

#include <stdio.h>
#include "nr.h"

#define N 20
#define NPTS 2000
#define ISCAL 200
#define LLEN 50

main()
{
    char words[LLEN+1];
    int i,ia,idum=(-13),j,k,klim,dist[N+1];
    float dd;

    for (;;) {
        for (j=0;j<=N;j++) dist[j]=0;
        do {
            printf("Select order of Gamma distribution (n=1..%d), \
            -1 to end\n",N);
            scanf("%d",&ia);
        } while (ia > N);
        if (ia < 0) break;
        for (i=1;i<=NPTS;i++) {
            j=gamdev(ia,&idum);
            if ((j >= 0) && (j <= N)) ++dist[j];
        }
        printf("\ngamma-distribution deviate, order %2d of %6d points\n",
            ia,NPTS);
        printf("%6s %7s %9s \n","x","p(x)","graph:");
        for (j=0;j<N;j++) {
            dd=(float) dist[j]/NPTS;
            for (k=1;k<=50;k++) words[k]=' ';
            klim=ISCAL*dd;
            if (klim > LLEN) klim=LLEN;
            for (k=1;k<=klim;k++) words[k]='*';
            printf("%6d %8.4f  ",j,dd);
            for (k=1;k<=klim;k++) printf("%c",words[k]);
            printf("\n");
        }
    }
}
/* Driver for routine POIDEV */
```

```
#include <stdio.h>
#include "nr.h"

#define N 20
#define NPTS 2000
#define ISCAL 200
#define LLEN 50

main()
{
    char txt[LLEN+1];
    int i,idum=(-13),j,k,klim,dist[N+1];
    float xm,dd;

    for (;;) {
        for (j=0;j<=N;j++) dist[j]=0;
        do {
            printf("Mean of distribution (0.0<x<%d.0) ",N);
            printf("- Negative to end:\n");
            scanf("%f",&xm);
        } while (xm > N);
        if (xm < 0.0) break;
        for (i=1;i<=NPTS;i++) {
            j=(int) (0.5+poidev(xm,&idum));
            if ((j >= 0) && (j <= N)) ++dist[j];
        }
        printf("Poisson-distributed deviate, mean %5.2f of %6d points\n",
            xm,NPTS);
        printf("%5s %8s %10s\n","x","p(x)","graph:");
        for (j=0;j<=N;j++) {
            dd=(float) dist[j]/NPTS;
            for (k=0;k<=LLEN;k++) txt[k]=' ';
            klim=ISCAL*dd;
            if (klim > LLEN) klim=LLEN;
            for (k=1;k<=klim;k++) txt[k]='*';
            txt[LLEN]='\0';
            printf("%6d %8.4f   %s\n",j,dd,txt);
        }
    }
}
/* Driver for routine BNLDEV */

#include <stdio.h>
#include "nr.h"

#define N 20
#define NPTS 2000
#define ISCAL 200
#define NN 100
#define LLEN 50

main()
{
    char txt[LLEN+1];
    int i,j,k,idum=(-133),klim,dist[N+1];
    float pp,xm,dd;
```

```
for (;;) {
    for (j=0;j<=N;j++) dist[j]=0;
    do {
        printf("Mean of distribution (0.0 to %d.0)",N);
        printf(" - Negative to end:\n");
        scanf("%f",&xm);
    } while (xm > 20.0);
    if (xm < 0.0) break;
    pp=xm/NN;
    for (i=1;i<=NPTS;i++) {
        j=bnldev(pp,NN,&idum);
        if (j >= 0 && j <= N) ++dist[j];
    }
    printf("Binomial-distributed deviate, mean %5.2f of %6d points\n",
        xm,NPTS);
    printf("%4s %8s %10s\n","x","p(x)","graph:");
    for (j=0;j<N;j++) {
        for (k=0;k<=LLEN;k++) txt[k]=' ';
        dd=(float) dist[j]/NPTS;
        klim=ISCAL*dd+1;
        if (klim > LLEN) klim=LLEN;
        for (k=1;k<=klim;k++) txt[k]='*';
        txt[LLEN]='\0';
        printf("%4d %9.4f    %s\n",j,dd,txt);
    }
}
}
```

Procedures `irbit1` and `irbit2` both generate random series of ones and zeros, using the bit-manipulation abilities of C. The sample programs `xirbit1.c` and `xir-bit2.c` for the two are the same, and they check that the series have correct statistical properties (or more exactly, that they have at least one correct property). They look for a 1 in the series and count how many zeros follow it before the next 1 appears. The result is stored as a distribution. There should be, for example, a 50% chance of no zeros, a 25% chance of exactly one zero, and so on.

```
/* Driver for routine IRBIT1 */

#include <stdio.h>
#include "nr.h"

#define NBIN 15
#define NTRIES 4000

main()
{
    int i,iflg,ipts=0,j,n;
    unsigned long int iseed=12345;
    float twoinv,delay[NBIN+1];

    /* Calculate distribution of runs of zeros */
    for (i=1;i<=NBIN;i++) delay[i]=0.0;
    printf("distribution of runs of n zeros\n");
    printf("%6s %22s %18s\n","n","probability","expected");
    for (i=1;i<=NTRIES;i++) {
        if (irbit1(&iseed) == 1) {
```

```
            ++ipts;
            iflg=0;
            for (j=1;j<=NBIN;j++) {
                if ((irbit1(&iseed) == 1) && (iflg == 0)) {
                    iflg=1;
                    delay[j] += 1.0;
                }
            }
        }
    }
    twoinv=0.5;
    for (n=1;n<=NBIN;n++) {
        printf("%6d %19.4f %20.4f\n",
            (n-1),delay[n]/ipts,twoinv);
        twoinv /= 2.0;
    }
}

/* Driver for routine IRBIT2 */

#include <stdio.h>
#include "nr.h"

#define NBIN 15
#define NTRIES 4000

float twoton(n)
int n;
{
    return n == 0 ? 1.0 : 2.0*twoton(n-1);
}

main()
{
    int i,iflg,ipts,j,n;
    unsigned long int iseed=111;
    float delay[NBIN+1];

    /* Calculate distribution of runs of zeros */
    for (i=1;i<=NBIN;i++) delay[i]=0.0;
    ipts=0;
    for (i=1;i<=NTRIES;i++) {
        if (irbit2(&iseed) == 1) {
            ++ipts;
            iflg=0;
            for (j=1;j<=NBIN;j++) {
                if ((irbit2(&iseed) == 1) && (iflg == 0)) {
                    iflg=1;
                    delay[j] += 1.0;
                }
            }
        }
    }
    printf("distribution of runs of n zeros\n");
    printf("%6s %22s %18s \n","n","probability","expected");
    for (n=1;n<=NBIN;n++)
        printf("%6d %19.4f %20.4f\n",
```

```
                       (n-1),delay[n]/ipts,1.0/twoton(n));
}
```

The next routine, ran4, is a random number generator with a uniform deviate, based on the data encryption standard des. When applied to ran4, the routine we used to demonstrate ran0 to ran3 is outrageously time consuming. ran4 is random but very slow. To try out ran4 we simply list the first ten random numbers for a given seed idum=-123. Compare your results to these; they should be exactly the same. We also generate 50 more random numbers and find their average and variance with avevar.

```
/* Driver for routine RAN4 */

#include <stdio.h>
#include "nr.h"
#include "nrutil.h"

#define NPT 50

main()
{
    int idum=(-123),j;
    float ave,vrnce,*y;

    y=vector(1,NPT);
    printf("\nFirst 10 Random Numbers with idum = %5d\n\n",idum);
    printf("%4s %11s \n","#","RAN4");
    for (j=1;j<=10;j++) {
        y[j]=ran4(&idum);
        printf("%4d %12.6f\n",j,y[j]);
    }
    printf("\nAverage and Variance of Next %3d\n",NPT);
    for (j=1;j<=NPT;j++)
        y[j]=ran4(&idum);
    avevar(y,NPT,&ave,&vrnce);
    printf("\nAverage: %10.4f\nVariance: %9.4f\n",ave,vrnce);
    printf("\nExpected Result for an Infinite Sample:\n");
    printf("\nAverage: %10.4f\nVariance: %9.4f\n",0.5,0.0833333);
    free_vector(y,1,NPT);
}
```

```
        First 10 random numbers with IDUM = -123
        1       .076597
        2       .533635
        3       .919756
        4       .317618
        5       .187471
        6       .629516
        7       .588766
        8       .953446
        9       .366207
        10      .915449
        Average and Variance of next   50
        Average:        .4542
        Variance:       .0786
```

des is a software implementation of the national data encryption standard. The complete formal test for this standard, though long and detailed, is included in program xdes.c. This test consists of feeding in a long series of input codes and checking the output for agreement with expected output codes. The input-output pairs comprising the test are contained in file destst.dat, listed in the Appendix to this chapter. To save you the job of comparing the many 16-character strings for accuracy, we have ended each line with the phrase "o.k." or "wrong", depending on the outcome.

The C-version of this program is quite different from the FORTRAN and Pascal versions, due to the availability of bit-manipulation operators. Here, the individual bits are packed into compact structures. The 64-bit quantities like 'key' and 'iin' are defined as a structure called 'immense' which contains two long integers. Likewise, 48-bit items are declared as 'huge', a structure containing three short integers. The order in which bits are read in from the NBS test file for DES is such that lowest order bits come first. Hence, we have included the function reverse which flips the bits end-to-end for input and output.

```
/* Driver for routine DES */

#include <stdio.h>
#include "nr.h"
#include "nrutil.h"

/* Converts character representing hexadecimal number to its integer
value in a machine-independent way. */
int hex2int(ch)
char ch;
{
    return ch >= '0' && ch <= '9' ? (int) (ch-'0') : (int) (ch-'A'+10);
}

/* Inverse of hex2int */
char int2hex(i)
int i;
{
    return i <= 9 ? (char) (i+'0') : (char) (i-10+'A');
}

/* Reverse bits of type immense */
void reverse(input)
immense *input;
{
    immense temp;
    int i;

    temp.r=temp.l=0L;
    for (i=1;i<=32;i++) {
        temp.r = (temp.r <<= 1) | ((*input).l & 1L);
        temp.l = (temp.l <<= 1) | ((*input).r & 1L);
        (*input).r >>= 1;
        (*input).l >>= 1;
    }
    (*input).r=temp.r;
    (*input).l=temp.l;
}

extern unsigned long bit[];
```

```
main()
{
    int i,idirec,j,m,mm,nciphr,newkey,strncmp();
    immense iin,iout,key;
    char hin[18],hkey[18],hout[18],hcmp[18];
    char verdct[8],txt[61],txt2[9],*strcpy();
    FILE *fp;

    if ((fp = fopen("destst.dat","r")) == NULL)
        nrerror("Data file DESTST.DAT not found\n");
    fgets(txt,60,fp);
    printf("\n%s",txt);
    for (;;) {
        fgets(txt,60,fp);
        printf("%s",txt);
        fscanf(fp,"%d %*s ",&nciphr);
        if (feof(fp)) break;
        fgets(txt2,8,fp);
        if (strncmp(txt2,"encode",6) == 0) idirec=0;
        if (strncmp(txt2,"decode",6) == 0) idirec=1;
        do {
            printf("%8s %20s %20s %15s\n","key","plaintext",
                "expected cipher","actual cipher");
            mm=16;
            if (nciphr < 16) mm=nciphr;
            nciphr -= 16;
            for (m=1;m<=mm;m++) {
                fscanf(fp,"%s %s %s ",hkey+1,hin+1,hcmp+1);
                iin.l=iin.r=key.l=key.r=0L;
                for (i=1,j=9;i<=8;i++,j++) {
                    iin.l=(iin.l <<= 4) | hex2int(hin[i]);
                    key.l=(key.l <<= 4) | hex2int(hkey[i]);
                    iin.r=(iin.r <<= 4) | hex2int(hin[j]);
                    key.r=(key.r <<= 4) | hex2int(hkey[j]);
                }
                newkey=1;
                reverse(&iin);
                reverse(&key);
                des(iin,key,&newkey,idirec,&iout);
                reverse(&iout);
                for (i=8,j=16;i>=1;i--,j--) {
                    hout[i]=int2hex((int) (iout.l & 0xf));
                    hout[j]=int2hex((int) (iout.r & 0xf));
                    iout.l >>= 4;
                    iout.r >>= 4;
                }
                hout[17]='\0';
                strncmp(hcmp+1,hout+1,16) == 0 ?
                    strcpy(verdct,"o.k.") :
                    strcpy(verdct,"wrong");
                printf("%16s %16s %16s %16s %s\n",
                    hkey+1,hin+1,hcmp+1,hout+1,verdct);
            }
            printf("Press RETURN to continue ...\n");
            getchar();
        } while (nciphr > 0);
    }
}
```

}

   desks contains two auxiliary routines for des, namely cyfun and ks. Technically, these were fully tested by the previous formal procedure. Sample program xdesks.c is simply an additional routine to help you track down problems. It first feeds a single key to ks and generates 16 subkeys. These subkeys are strings of zeros and ones which we print out as strings of "-" and "*". In this form the results are easier to compare with printed text. Next an input vector is fed to the cipher function cyfun along with each of these subkeys. The results are again printed in "-" and "*" characters. If all is well, you will observe these patterns:

```
Legend:
                    -=0    *=1
Master key:
       *-*-*-*-*-*-*-*-*-*-*-*-*-*-*-*-*-*-*-*-*-*-*-*-*-*-*-*-*-
Sub-master keys:
       1     -*--****-**-**-*--**-*-*---*-*-*-*****-**-*-*-**
       2     -*--*******--*-*--*-**-*--*-***--****-**-*-*-**
       3     **--*-***----*-**-*****-*-***--*-**--*-***--**
       4     *****--**---*-*-*-*-****----*****--*-**-***--*-
       5     *-**--**-***-*-*-*-*-**-*-*-**---****-*-**---
       6     *-**------*****-**-*-**-**-**--**--*-**-*-***--
       7     -***-*---******--*-*-*---*-**--*****-**-*-*-**--
       8     -*---**-****-*-*-***-*----***----****--*-*-**-*
       9     **---**-***--*-*-***-*-*--***-*--*****---*-**-*-*
      10     **--*****---***--**--***-*-*-**-**-*--**-**--**
      11     ***-*****--*--***-*-***-*---***-**-*-**---*---**
      12     *-***-***--*--*-**--*-*****-***------**-*-*-**-
      13     --***--*-*-**-*-**-**-**-*-**-*-*-**-----****--***-
      14     --**-*---****--***-***---*-*-*--*-**-**-**--**-*
      15     ---*-**--**-**-*-*-*-*-*-****-*-*-**-*--***-**-*
      16     -*-****--**-**-*-*-*-*-*-**-*-*-****-*---*-*-****
Legend:
                    -=0    *=1
Input to cipher function:
       *--*--*--*--*--*--*--*--*--*-
Ciphered output:
       1     -*----**-*--------****-----***-*
       2     --*--***-**--*-*--****-*-*----*-*
       3     --*****--*-*--*--**---*----***--*
       4     -**-*--*---*--**---***-****---*-*
       5     *-*---*---****---*-*-*--*-*-**--
       6     --*--**--*---**--**---*------*
       7     *-*-****-*-**-*-------*-*----****
       8     -***--***----**-*-----*---*--**-
       9     -**-*-*****-***-****---------*--
      10     *--***--**-******-*--**-*-***-*-
      11     -*****-*----**--*****-*-**-*---
      12     ***-*-----***---**---*--*-*-*-*-
      13     -**-*-----*-****-*----------**-*
      14     ----*---**-***----*-*----**------
      15     *-**--*-*-**--**-*-*--***-*-****
      16     *---*--*---*-*-********-***-*-***-
```

/* Driver for routines KS and CYFUN in file DESKS.C */

#include <stdio.h>

```
#include "nr.h"
#include "nrutil.h"

unsigned long bit[33];

main()
{
    int i,k,l,m;
    char txt[65];
    unsigned long iout,ir;
    immense key;
    great kn;

    bit[1]=1;
    for (i=2;i<=32;i++) bit[i]=bit[i-1] << 1;
    /* Test routine KS */
    key.l=key.r=0x55555555L;
    for (i=1;i<=64;i++) txt[i]= i%2 ? '*' : '-';
    printf("legend:\n%25s\n","-=0    *=1 ");
    printf("master key:\n                            ");
    for (i=1;i<=64;i++) printf("%1c",txt[i]);
    printf("\nsub-master keys:\n");
    for (i=1;i<=16;i++) {
        ks(key,i,&kn);
        for (k=1,l=17,m=33;k<=16;k++,l++,m++) {
            txt[k]=kn.r & bit[k] ? '*' : '-';
            txt[l]=kn.c & bit[k] ? '*' : '-';
            txt[m]=kn.l & bit[k] ? '*' : '-';
        }
        printf("%6d %6s",i," ");
        for (k=1;k<=48;k++) printf("%1c",txt[k]);
        printf("\n");
    }
    printf("Press RETURN to continue...\n");
    getchar();
    /* Test routine CYFUN */
    ir=0x49249249L;
    for (i=1;i<=32;i++)
        txt[i]=(i%3)%2 ? '*' : '-';
    printf("legend:\n%25s\n","-=0    *=1 ");
    printf("input to cipher function:\n                    ");
    for (i=1;i<=32;i++) printf("%1c",txt[i]);
    printf("\nciphered output:\n");
    for (i=1;i<=16;i++) {
        ks(key,i,&kn);
        cyfun(ir,kn,&iout);
        for (k=1;k<=32;k++) txt[k]=iout & bit[k] ? '*' : '-';
        printf("%6d %6s",i," ");
        for (k=1;k<=32;k++) printf("%1c",txt[k]);
        printf("\n");
    }
}
```

Appendix

File destst.dat:

```
DES Validation, as per NBS publication 500-20
*** Initial Permutation and Expansion test: ***
        Key             Plaintext       Expected Cipher
   0101010101010101 95F8A5E5DD31D900 8000000000000000
   0101010101010101 DD7F121CA5015619 4000000000000000
   0101010101010101 2E8653104F3834EA 2000000000000000
   0101010101010101 4BD388FF6CD81D4F 1000000000000000
   0101010101010101 20B9E767B2FB1456 0800000000000000
   0101010101010101 55579380D77138EF 0400000000000000
   0101010101010101 6CC5DEFAAF04512F 0200000000000000
   0101010101010101 0D9F279BA5D87260 0100000000000000
   0101010101010101 D9031B0271BD5A0A 0080000000000000
   0101010101010101 424250B37C3DD951 0040000000000000
   0101010101010101 B8061B7ECD9A21E5 0020000000000000
   0101010101010101 F15D0F286B65BD28 0010000000000000
   0101010101010101 ADD0CC8D6E5DEBA1 0008000000000000
   0101010101010101 E6D5F82752AD63D1 0004000000000000
   0101010101010101 ECBFE3BD3F591A5E 0002000000000000
   0101010101010101 F356834379D165CD 0001000000000000
   0101010101010101 2B9F982F20037FA9 0000800000000000
   0101010101010101 889DE068A16F0BE6 0000400000000000
   0101010101010101 E19E275D846A1298 0000200000000000
   0101010101010101 329A8ED523D71AEC 0000100000000000
   0101010101010101 E7FCE22557D23C97 0000080000000000
   0101010101010101 12A9F5817FF2D65D 0000040000000000
   0101010101010101 A484C3AD38DC9C19 0000020000000000
   0101010101010101 FBE00A8A1EF8AD72 0000010000000000
   0101010101010101 750D079407521363 0000008000000000
   0101010101010101 64FEED9C724C2FAF 0000004000000000
   0101010101010101 F02B263B328E2B60 0000002000000000
   0101010101010101 9D64555A9A10B852 0000001000000000
   0101010101010101 D106FF0BED5255D7 0000000800000000
   0101010101010101 E1652C6B138C64A5 0000000400000000
   0101010101010101 E428581186EC8F46 0000000200000000
   0101010101010101 AEB5F5EDE22D1A36 0000000100000000
   0101010101010101 E943D7568AEC0C5C 0000000080000000
   0101010101010101 DF98C8276F54B04B 0000000040000000
   0101010101010101 B160E4680F6C696F 0000000020000000
   0101010101010101 FA0752B07D9C4AB8 0000000010000000
   0101010101010101 CA3A2B036DBC8502 0000000008000000
   0101010101010101 5E0905517BB59BCF 0000000004000000
   0101010101010101 814EEB3B91D90726 0000000002000000
   0101010101010101 4D49DB1532919C9F 0000000001000000
   0101010101010101 25EB5FC3F8CF0621 0000000000800000
   0101010101010101 AB6A20C0620D1C6F 0000000000400000
   0101010101010101 79E90DBC98F92CCA 0000000000200000
   0101010101010101 866ECEDD8072BB0E 0000000000100000
   0101010101010101 8B54536F2F3E64A8 0000000000080000
   0101010101010101 EA51D3975595B86B 0000000000040000
   0101010101010101 CAFFC6AC4542DE31 0000000000020000
   0101010101010101 8DD45A2DDF90796C 0000000000010000
   0101010101010101 1029D55E880EC2D0 0000000000008000
   0101010101010101 5D86CB23639DBEA9 0000000000004000
```

```
0101010101010101 1D1CA853AE7C0C5F 0000000000002000
0101010101010101 CE332329248F3228 0000000000001000
0101010101010101 8405D1ABE24FB942 0000000000000800
0101010101010101 E643D78090CA4207 0000000000000400
0101010101010101 48221B9937748A23 0000000000000200
0101010101010101 DD7C0BBD61FAFD54 0000000000000100
0101010101010101 2FBC291A570DB5C4 0000000000000080
0101010101010101 E07C30D7E4E26E12 0000000000000040
0101010101010101 0953E2258E8E90A1 0000000000000020
0101010101010101 5B711BC4CEEBF2EE 0000000000000010
0101010101010101 CC083F1E6D9E85F6 0000000000000008
0101010101010101 D2FD8867D50D2DFE 0000000000000004
0101010101010101 06E7EA22CE92708F 0000000000000002
0101010101010101 166B40B44ABA4BD6 0000000000000001
*** Inverse Permutation and Expansion test ***
     Key           Plaintext         Expected Cipher
0101010101010101 8000000000000000 95F8A5E5DD31D900
0101010101010101 4000000000000000 DD7F121CA5015619
0101010101010101 2000000000000000 2E8653104F3834EA
0101010101010101 1000000000000000 4BD388FF6CD81D4F
0101010101010101 0800000000000000 20B9E767B2FB1456
0101010101010101 0400000000000000 55579380D77138EF
0101010101010101 0200000000000000 6CC5DEFAAF04512F
0101010101010101 0100000000000000 0D9F279BA5D87260
0101010101010101 0080000000000000 D9031B0271BD5A0A
0101010101010101 0040000000000000 424250B37C3DD951
0101010101010101 0020000000000000 B8061B7ECD9A21E5
0101010101010101 0010000000000000 F15D0F286B65BD28
0101010101010101 0008000000000000 ADD0CC8D6E5DEBA1
0101010101010101 0004000000000000 E6D5F82752AD63D1
0101010101010101 0002000000000000 ECBFE3BD3F591A5E
0101010101010101 0001000000000000 F356834379D165CD
0101010101010101 0000800000000000 2B9F982F20037FA9
0101010101010101 0000400000000000 889DE068A16F0BE6
0101010101010101 0000200000000000 E19E275D846A1298
0101010101010101 0000100000000000 329A8ED523D71AEC
0101010101010101 0000080000000000 E7FCE22557D23C97
0101010101010101 0000040000000000 12A9F5817FF2D65D
0101010101010101 0000020000000000 A484C3AD38DC9C19
0101010101010101 0000010000000000 FBE00A8A1EF8AD72
0101010101010101 0000008000000000 750D079407521363
0101010101010101 0000004000000000 64FEED9C724C2FAF
0101010101010101 0000002000000000 F02B263B328E2B60
0101010101010101 0000001000000000 9D64555A9A10B852
0101010101010101 0000000800000000 D106FF0BED5255D7
0101010101010101 0000000400000000 E1652C6B138C64A5
0101010101010101 0000000200000000 E428581186EC8F46
0101010101010101 0000000100000000 AEB5F5EDE22D1A36
0101010101010101 0000000080000000 E943D7568AEC0C5C
0101010101010101 0000000040000000 DF98C8276F54B04B
0101010101010101 0000000020000000 B160E4680F6C696F
0101010101010101 0000000010000000 FA0752B07D9C4AB8
0101010101010101 0000000008000000 CA3A2B036DBC8502
0101010101010101 0000000004000000 5E0905517BB59BCF
0101010101010101 0000000002000000 814EEB3B91D90726
0101010101010101 0000000001000000 4D49DB1532919C9F
0101010101010101 0000000000800000 25EB5FC3F8CF0621
```

```
0101010101010101  0000000000400000  AB6A20C0620D1C6F
0101010101010101  0000000000200000  79E90DBC98F92CCA
0101010101010101  0000000000100000  866ECEDD8072BB0E
0101010101010101  0000000000080000  8B54536F2F3E64A8
0101010101010101  0000000000040000  EA51D3975595B86B
0101010101010101  0000000000020000  CAFFC6AC4542DE31
0101010101010101  0000000000010000  8DD45A2DDF90796C
0101010101010101  0000000000008000  1029D55E880EC2D0
0101010101010101  0000000000004000  5D86CB23639DBEA9
0101010101010101  0000000000002000  1D1CA853AE7C0C5F
0101010101010101  0000000000001000  CE332329248F3228
0101010101010101  0000000000000800  8405D1ABE24FB942
0101010101010101  0000000000000400  E643D78090CA4207
0101010101010101  0000000000000200  48221B9937748A23
0101010101010101  0000000000000100  DD7C0BBD61FAFD54
0101010101010101  0000000000000080  2FBC291A570DB5C4
0101010101010101  0000000000000040  E07C30D7E4E26E12
0101010101010101  0000000000000020  0953E2258E8E90A1
0101010101010101  0000000000000010  5B711BC4CEEBF2EE
0101010101010101  0000000000000008  CC083F1E6D9E85F6
0101010101010101  0000000000000004  D2FD8867D50D2DFE
0101010101010101  0000000000000002  06E7EA22CE92708F
0101010101010101  0000000000000001  166B40B44ABA4BD6
*** Key Permutation tests: ***
      Key            Plaintext        Expected Cipher
8001010101010101  0000000000000000  95A8D72813DAA94D
4001010101010101  0000000000000000  0EEC1487DD8C26D5
2001010101010101  0000000000000000  7AD16FFB79C45926
1001010101010101  0000000000000000  D3746294CA6A6CF3
0801010101010101  0000000000000000  809F5F873C1FD761
0401010101010101  0000000000000000  C02FAFFEC989D1FC
0201010101010101  0000000000000000  4615AA1D33E72F10
0180010101010101  0000000000000000  2055123350C00858
0140010101010101  0000000000000000  DF3B99D6577397C8
0120010101010101  0000000000000000  31FE17369B5288C9
0110010101010101  0000000000000000  DFDD3CC64DAE1642
0108010101010101  0000000000000000  178C83CE2B399D94
0104010101010101  0000000000000000  50F636324A9B7F80
0102010101010101  0000000000000000  A8468EE3BC18F06D
0101800101010101  0000000000000000  A2DC9E92FD3CDE92
0101400101010101  0000000000000000  CAC09F797D031287
0101200101010101  0000000000000000  90BA680B22AEB525
0101100101010101  0000000000000000  CE7A24F350E280B6
0101080101010101  0000000000000000  882BFF0AA01A0B87
0101040101010101  0000000000000000  25610288924511C2
0101020101010101  0000000000000000  C71516C29C75D170
0101018001010101  0000000000000000  5199C29A52C9F059
0101014001010101  0000000000000000  C22F0A294A71F29F
0101012001010101  0000000000000000  EE371483714C02EA
0101011001010101  0000000000000000  A81FBD448F9E522F
0101010801010101  0000000000000000  4F644C92E192DFED
0101010401010101  0000000000000000  1AFA9A66A6DF92AE
0101010201010101  0000000000000000  B3C1CC715CB879D8
0101010180010101  0000000000000000  19D032E64AB0BD8B
0101010140010101  0000000000000000  3CFAA7A7DC8720DC
0101010120010101  0000000000000000  B7265F7F447AC6F3
0101010110010101  0000000000000000  9DB73B3C0D163F54
```

```
0101010108010101  0000000000000000  8181B65BABF4A975
0101010104010101  0000000000000000  93C9B64042EAA240
0101010102010101  0000000000000000  5570530829705592
0101010101800101  0000000000000000  8638809E878787A0
0101010101400101  0000000000000000  41B9A79AF79AC208
0101010101200101  0000000000000000  7A9BE42F2009A892
0101010101100101  0000000000000000  29038D56BA6D2745
0101010101080101  0000000000000000  5495C6ABF1E5DF51
0101010101040101  0000000000000000  AE13DBD561488933
0101010101020101  0000000000000000  024D1FFA8904E389
0101010101018001  0000000000000000  D1399712F99BF02E
0101010101014001  0000000000000000  14C1D7C1CFFEC79E
0101010101012001  0000000000000000  1DE5279DAE3BED6F
0101010101011001  0000000000000000  E941A33F85501303
0101010101010801  0000000000000000  DA99DBBC9A03F379
0101010101010401  0000000000000000  B7FC92F91D8E92E9
0101010101010201  0000000000000000  AE8E5CAA3CA04E85
0101010101010180  0000000000000000  9CC62DF43B6EED74
0101010101010140  0000000000000000  D863DBB5C59A91A0
0101010101010120  0000000000000000  A1AB2190545B91D7
0101010101010110  0000000000000000  0875041E64C570F7
0101010101010108  0000000000000000  5A594528BEBEF1CC
0101010101010104  0000000000000000  FCDB3291DE21F0C0
0101010101010102  0000000000000000  869EFD7F9F265A09
*** Test of right-shifts in Decryption ***
      Key             Plaintext       Expected Cipher
8001010101010101  95A8D72813DAA94D  0000000000000000
4001010101010101  0EEC1487DD8C26D5  0000000000000000
2001010101010101  7AD16FFB79C45926  0000000000000000
1001010101010101  D3746294CA6A6CF3  0000000000000000
0801010101010101  809F5F873C1FD761  0000000000000000
0401010101010101  C02FAFFEC989D1FC  0000000000000000
0201010101010101  4615AA1D33E72F10  0000000000000000
0180010101010101  2055123350C00858  0000000000000000
0140010101010101  DF3B99D6577397C8  0000000000000000
0120010101010101  31FE17369B5288C9  0000000000000000
0110010101010101  DFDD3CC64DAE1642  0000000000000000
0108010101010101  178C83CE2B399D94  0000000000000000
0104010101010101  50F636324A9B7F80  0000000000000000
0102010101010101  A8468EE3BC18F06D  0000000000000000
0101800101010101  A2DC9E92FD3CDE92  0000000000000000
0101400101010101  CAC09F797D031287  0000000000000000
0101200101010101  90BA680B22AEB525  0000000000000000
0101100101010101  CE7A24F350E280B6  0000000000000000
0101080101010101  882BFF0AA01A0B87  0000000000000000
0101040101010101  25610288924511C2  0000000000000000
0101020101010101  C71516C29C75D170  0000000000000000
0101018001010101  5199C29A52C9F059  0000000000000000
0101014001010101  C22F0A294A71F29F  0000000000000000
0101012001010101  EE371483714C02EA  0000000000000000
0101011001010101  A81FBD448F9E522F  0000000000000000
0101010801010101  4F644C92E192DFED  0000000000000000
0101010401010101  1AFA9A66A6DF92AE  0000000000000000
0101010201010101  B3C1CC715CB879D8  0000000000000000
0101010180010101  19D032E64AB0BD8B  0000000000000000
0101010140010101  3CFAA7A7DC8720DC  0000000000000000
0101010120010101  B7265F7F447AC6F3  0000000000000000
```

```
0101010110010101  9DB73B3C0D163F54  0000000000000000
0101010108010101  8181B65BABF4A975  0000000000000000
0101010104010101  93C9B64042EAA240  0000000000000000
0101010102010101  5570530829705592  0000000000000000
0101010101800101  8638809E878787A0  0000000000000000
0101010101400101  41B9A79AF79AC208  0000000000000000
0101010101200101  7A9BE42F2009A892  0000000000000000
0101010101100101  29038D56BA6D2745  0000000000000000
0101010101080101  5495C6ABF1E5DF51  0000000000000000
0101010101040101  AE13DBD561488933  0000000000000000
0101010101020101  024D1FFA8904E389  0000000000000000
0101010101018001  D1399712F99BF02E  0000000000000000
0101010101014001  14C1D7C1CFFEC79E  0000000000000000
0101010101012001  1DE5279DAE3BED6F  0000000000000000
0101010101011001  E941A33F85501303  0000000000000000
0101010101010801  DA99DBBC9A03F379  0000000000000000
0101010101010401  B7FC92F91D8E92E9  0000000000000000
0101010101010201  AE8E5CAA3CA04E85  0000000000000000
0101010101010180  9CC62DF43B6EED74  0000000000000000
0101010101010140  D863DBB5C59A91A0  0000000000000000
0101010101010120  A1AB2190545B91D7  0000000000000000
0101010101010110  0875041E64C570F7  0000000000000000
0101010101010108  5A594528BEBEF1CC  0000000000000000
0101010101010104  FCDB3291DE21F0C0  0000000000000000
0101010101010102  869EFD7F9F265A09  0000000000000000
```

*** Data permutation test: ***

Key	Plaintext	Expected Cipher
1046913489980131	0000000000000000	88D55E54F54C97B4
1007103489988020	0000000000000000	0C0CC00C83EA48FD
10071034C8980120	0000000000000000	83BC8EF3A6570183
1046103489988020	0000000000000000	DF725DCAD94EA2E9
1086911519190101	0000000000000000	E652B53B550BE8B0
1086911519580101	0000000000000000	AF527120C485CBB0
5107B01519580101	0000000000000000	0F04CE393DB926D5
1007B01519790101	0000000000000000	C9F00FFC74079067
3107915498080101	0000000000000000	7CFD82A593252B4E
3107919498080101	0000000000000000	CB49A2F9E91363E3
10079115B9080140	0000000000000000	00B588BE70D23F56
3107911598080140	0000000000000000	406A9A6AB43399AE
1007D01589980101	0000000000000000	6CB773611DCA9ADA
9107911589980101	0000000000000000	67FD21C17DBB5D70
9107D01589190101	0000000000000000	9592CB4110430787
1007D01598980120	0000000000000000	A6B7FF68A318DDD3
1007940498190101	0000000000000000	4D102196C914CA16
0107910491190401	0000000000000000	2DFA9F4573594965
0107910491190101	0000000000000000	B46604816C0E0774
0107940491190401	0000000000000000	6E7E6221A4F34E87
19079210981A0101	0000000000000000	AA85E74643233199
1007911998190801	0000000000000000	2E5A19DB4D1962D6
10079119981A0801	0000000000000000	23A866A809D30894
1007921098190101	0000000000000000	D812D961F017D320
100791159819010B	0000000000000000	055605816E58608F
1004801598190101	0000000000000000	ABD88E8B1B7716F1
1004801598190102	0000000000000000	537AC95BE69DA1E1
1004801598190108	0000000000000000	AED0F6AE3C25CDD8
1002911498100104	0000000000000000	B3E35A5EE53E7B8D
1002911598190104	0000000000000000	61C79C71921A2EF8

```
1002911598100201  0000000000000000  E2F5728F0995013C
1002911698100101  0000000000000000  1AEAC39A61F0A464
*** S-Box test: ***
       Key            Plaintext        Expected Cipher
7CA110454A1A6E57  01A1D6D039776742  690F5B0D9A26939B
0131D9619DC1376E  5CD54CA83DEF57DA  7A389D10354BD271
07A1133E4A0B2686  0248D43806F67172  868EBB51CAB4599A
3849674C2602319E  51454B582DDF440A  7178876E01F19B2A
04B915BA43FEB5B6  42FD443059577FA2  AF37FB421F8C4095
0113B970FD34F2CE  059B5E0851CF143A  86A560F10EC6D85B
0170F175468FB5E6  0756D8E0774761D2  0CD3DA020021DC09
43297FAD38E373FE  762514B829BF486A  EA676B2CB7DB2B7A
07A7137045DA2A16  3BDD119049372802  DFD64A815CAF1A0F
04689104C2FD3B2F  26955F6835AF609A  5C513C9C4886C088
37D06BB516CB7546  164D5E404F275232  0A2AEEAE3FF4AB77
1F08260D1AC2465E  6B056E18759F5CCA  EF1BF03E5DFA575A
584023641ABA6176  004BD6EF09176062  88BF0DB6D70DEE56
025816164629B007  480D39006EE762F2  A1F9915541020B56
49793EBC79B3258F  437540C8698F3CFA  6FBF1CAFCFFD0556
4FB05E1515AB73A7  072D43A077075292  2F22E49BAB7CA1AC
49E95D6D4CA229BF  02FE55778117F12A  5A6B612CC26CCE4A
018310DC409B26D6  1D9D5C5018F728C2  5F4C038ED12B2E41
1C587F1C13924FEF  305532286D6F295A  63FAC0D034D9F793
*** End of Test ***
```

Chapter 8: Sorting

Chapter 8 of *Numerical Recipes* covers a variety of sorting tasks including sorting arrays into numerical order, preparing an index table for the order of an array, and preparing a rank table showing the rank order of each element in the array. `piksrt` sorts a single array by the straight insertion method. `piksr2` sorts by the same method but makes the corresponding rearrangement of a second array as well. `shell` carries out a *Shell sort*. `sort` and `sort2` both do a *Heapsort*, and they are related in the same way as `piksrt` and `piksr2`. That is, `sort` sorts a single array; `sort2` sorts an array while correspondingly rearranging a second array. `qcksrt` sorts an array by the *Quicksort* algorithm, which is fast (on average) but requires a small amount of auxiliary storage.

`indexx` indexes an array. That is, it produces a second array that references the elements of the original array in the order of their size. `sort3` uses `indexx` and illustrates its value by sorting one array while making corresponding rearrangements in two others. `rank` produces the rank table for an array of data. The rank table is a second array whose elements list the rank order of the corresponding elements of the original array.

Finally, the routines `eclass` and `eclazz` deal with equivalence classes. `eclass` gives the equivalence class of each element in an array based on a list of equivalent pairs which it is given as input. `eclazz` gives the same output but bases it on a procedure named `equiv(j,k)` which tells whether two array elements `j` and `k` are in the same equivalence class.

★   ★   ★   ★

Routine `piksrt` sorts an array by straight insertion. Sample program `xpiksrt.c` provides it with a 100-element array from file `tarray.dat` which is listed in the Appendix to this chapter. The program prints both the original and the sorted array for comparison.

```
/* Driver for routine PIKSRT */

#include <stdio.h>
#include "nr.h"
#include "nrutil.h"

#define MAXSTR 80
#define NP 100

main()
{
```

102

```
      char txt[MAXSTR];
      int i,j;
      float *a;
      FILE *fp;

      a=vector(1,NP);
      if ((fp = fopen("tarray.dat","r")) == NULL)
          nrerror("Data file TARRAY.DAT not found\n");
      fgets(txt,MAXSTR,fp);
      for (i=1;i<=NP;i++) fscanf(fp,"%f",&a[i]);
      fclose(fp);
      printf("original array:\n");
      for (i=0;i<=9;i++) {
          for (j=1;j<=10;j++) printf("%7.2f",a[10*i+j]);
          printf("\n");
      }
      piksrt(NP,a);
      printf("sorted array:\n");
      for (i=0;i<=9;i++) {
          for (j=1;j<=10;j++) printf("%7.2f",a[10*i+j]);
          printf("\n");
      }
      free_vector(a,1,NP);
}
```

piksr2 sorts an array, and simultaneously rearranges a second array (of the same size) correspondingly. In program xpiksr2.c, the first array a[i] is again taken from tarray.dat. The second is defined by b[i]=i-1. In other words, b is originally sorted and a is not. After a call to piksr2, the situation should be reversed. With a second call, this time with b as the first argument and a as the second, the two arrays should be returned to their original form.

```
/* Driver for routine PIKSR2 */

#include <stdio.h>
#include "nr.h"
#include "nrutil.h"

#define NP 100
#define MAXSTR 80

main()
{
      char txt[MAXSTR];
      int i,j;
      float *a,*b;
      FILE *fp;

      a=vector(1,NP);
      b=vector(1,NP);
      if ((fp = fopen("tarray.dat","r")) == NULL)
          nrerror("Data file TARRAY.DAT not found\n");
      fgets(txt,MAXSTR,fp);
      for (i=1;i<=NP;i++) fscanf(fp,"%f",&a[i]);
      fclose(fp);
      /* generate b-array */
      for (i=1;i<=NP;i++) b[i]=i-1;
```

```
    /* sort a and mix b */
    piksr2(NP,a,b);
    printf("\nAfter sorting a and mixing b, array a is:\n");
    for (i=0;i<=9;i++) {
        for (j=1;j<=10;j++) printf("%7.2f",a[10*i+j]);
        printf("\n");
    }
    printf("\n... and array b is:\n");
    for (i=0;i<=9;i++) {
        for (j=1;j<=10;j++) printf("%7.2f",b[10*i+j]);
        printf("\n");
    }
    printf("press return to continue ...\n");
    getchar();
    /* sort b and mix a */
    piksr2(NP,b,a);
    printf("\nAfter sorting b and mixing a, array a is:\n");
    for (i=0;i<=9;i++) {
        for (j=1;j<=10;j++) printf("%7.2f",a[10*i+j]);
        printf("\n");
    }
    printf("\n... and array b is:\n");
    for (i=0;i<=9;i++) {
        for (j=1;j<=10;j++) printf("%7.2f",b[10*i+j]);
        printf("\n");
    }
    free_vector(b,1,NP);
    free_vector(a,1,NP);
}
```

Procedure `shell` does a Shell sort of a data array. The calling format is identical to that of `piksrt`, and so we use the same sample program, now called `xshell.c`.

```
/* Driver for routine SHELL */

#include <stdio.h>
#include "nr.h"
#include "nrutil.h"

#define MAXSTR 80
#define NP 100

main()
{
    char txt[MAXSTR];
    int i,j;
    float *a;
    FILE *fp;

    a=vector(1,NP);
    if ((fp = fopen("tarray.dat","r")) == NULL)
        nrerror("Data file TARRAY.DAT not found\n");
    fgets(txt,MAXSTR,fp);
    for (i=1;i<=NP;i++) fscanf(fp,"%f",&a[i]);
    fclose(fp);
    printf("\nOriginal array:\n");
    for (i=0;i<=9;i++) {
        for (j=1;j<=10;j++) printf("%7.2f",a[10*i+j]);
```

```
        printf("\n");
    }
    shell(NP,a);
    printf("\nSorted array:\n");
    for (i=0;i<=9;i++) {
        for (j=1;j<=10;j++) printf("%7.2f",a[10*i+j]);
        printf("\n");
    }
    free_vector(a,1,NP);
}
```

By the same token, routines `sort` and `sort2` employ the same programs as routines `piksrt` and `piksr2`, respectively. Both routines use the Heapsort algorithm. `sort`, however, works on a single array. `sort2` sorts one array while making corresponding rearrangements to a second.

```
/* Driver for routine SORT */

#include <stdio.h>
#include "nr.h"
#include "nrutil.h"

#define MAXSTR 80
#define NP 100

main()
{
    char txt[MAXSTR];
    int i,j;
    float *a;
    FILE *fp;

    a=vector(1,NP);
    if ((fp = fopen("tarray.dat","r")) == NULL)
        nrerror("Data file TARRAY.DAT not found\n");
    fgets(txt,MAXSTR,fp);
    for (i=1;i<=NP;i++) fscanf(fp,"%f",&a[i]);
    fclose(fp);
    printf("\noriginal array:\n");
    for (i=0;i<=9;i++) {
        for (j=1;j<=10;j++) printf("%7.2f",a[10*i+j]);
        printf("\n");
    }
    sort(NP,a);
    printf("\nsorted array:\n");
    for (i=0;i<=9;i++) {
        for (j=1;j<=10;j++) printf("%7.2f",a[10*i+j]);
        printf("\n");
    }
    free_vector(a,1,NP);
}
/* Driver for routine SORT2 */

#include <stdio.h>
#include "nr.h"
#include "nrutil.h"
```

```
#define MAXSTR 80
#define NP 100

main()
{
    char txt[MAXSTR];
    int i,j;
    float *a,*b;
    FILE *fp;

    a=vector(1,NP);
    b=vector(1,NP);
    if ((fp = fopen("tarray.dat","r")) == NULL)
        nrerror("Data file TARRAY.DAT not found\n");
    fgets(txt,MAXSTR,fp);
    for (i=1;i<=NP;i++) fscanf(fp,"%f",&a[i]);
    fclose(fp);
    /* generate b-array */
    for (i=1;i<=NP;i++) b[i]=i-1;
    /* sort a and mix b */
    sort2(NP,a,b);
    printf("\nAfter sorting a and mixing b, array a is:\n");
    for (i=0;i<=9;i++) {
        for (j=1;j<=10;j++) printf("%7.2f",a[10*i+j]);
        printf("\n");
    }
    printf("\n... and array b is:\n");
    for (i=0;i<=9;i++) {
        for (j=1;j<=10;j++) printf("%7.2f",b[10*i+j]);
        printf("\n");
    }
    printf("press return to continue...\n");
    getchar();
    /* sort b and mix a */
    sort2(NP,b,a);
    printf("\nAfter sorting b and mixing a, array a is:\n");
    for (i=0;i<=9;i++) {
        for (j=1;j<=10;j++) printf("%7.2f",a[10*i+j]);
        printf("\n");
    }
    printf("\n... and array b is:\n");
    for (i=0;i<=9;i++) {
        for (j=1;j<=10;j++) printf("%7.2f",b[10*i+j]);
        printf("\n");
    }
    free_vector(b,1,NP);
    free_vector(a,1,NP);
}
```

The procedure `indexx` generates the index array for a given input array. The index array `indx[j]` gives, for each `j`, the index of the element of the input array which will assume position `j` if the array is sorted. That is, for an input array a, the sorted version of a will be `a[indx[j]]`. To demonstrate this, sample program `xindexx.c` produces an index for the array in `tarray.dat`. It then prints the array in the order `a[indx[j]]`, $j=1, .., 100$ for inspection.

```
/* Driver for routine INDEXX */

#include <stdio.h>
#include "nr.h"
#include "nrutil.h"

#define NP 100
#define MAXSTR 80

main()
{
    char txt[MAXSTR];
    int i,j,*indx;
    float *a;
    FILE *fp;

    indx=ivector(1,NP);
    a=vector(1,NP);
    if ((fp = fopen("tarray.dat","r")) == NULL)
        nrerror("Data file TARRAY.DAT not found\n");
    fgets(txt,MAXSTR,fp);
    for (i=1;i<=NP;i++) fscanf(fp,"%f",&a[i]);
    fclose(fp);
    indexx(NP,a,indx);
    printf("\noriginal array:\n");
    for (i=0;i<=9;i++) {
        for (j=1;j<=10;j++) printf("%7.2f",a[10*i+j]);
        printf("\n");
    }
    printf("\nsorted array:\n");
    for (i=0;i<=9;i++) {
        for (j=1;j<=10;j++) printf("%7.2f",a[indx[10*i+j]]);
        printf("\n");
    }
    free_vector(a,1,NP);
    free_ivector(indx,1,NP);
}
```

One use for `indexx` is the management of more than two arrays. `sort3`, for example, sorts one array while making corresponding reorderings of two other arrays. In sample program `xsort3.c`, the first array is taken as the first 64 elements of `tarray.dat` (see Appendix). The second and third arrays are taken to be the numbers 1 to 64 in forward order and reverse order, respectively. When the first array is ordered, the second and third are scrambled, but scrambled in exactly the same way. To prove this, a text message is assigned to a character array. Then the letters are scrambled according to the order of numbers found in the rearranged second array. They are subsequently unscrambled according to the order of numbers found in the rearranged third array. If `sort3` works properly, this ought to leave the message reading in the reverse order.

```
/* Driver for routine SORT3 */

#include <stdio.h>
#include <math.h>
#include "nr.h"
#include "nrutil.h"
```

```
#define NLEN 64

main()
{
    int i,j;
    char dummy[NLEN],amsg[NLEN+1],bmsg[NLEN+1],cmsg[NLEN+1];
    char *strcpy(),*strcat();
    float *a,*b,*c;
    FILE *fp;

    a=vector(1,NLEN);
    b=vector(1,NLEN);
    c=vector(1,NLEN);
    (void) strcpy(amsg,"I'd rather have a bottle in front of");
    (void) strcat(amsg," me than a frontal lobotomy.");
    printf("\noriginal message:\n%s\n",amsg);
    /* read array of random numbers */
    if ((fp = fopen("tarray.dat","r")) == NULL)
        nrerror("Data file TARRAY.DAT not found\n");
    fgets(dummy,NLEN,fp);
    for (i=1;i<=NLEN;i++) fscanf(fp,"%f",&a[i]);
    fclose(fp);
    /* create array b and array c */
    for (i=1;i<=NLEN;i++) {
        b[i]=i;
        c[i]=NLEN+1-i;
    }
    /* sort array a while mixing b and c */
    sort3(NLEN,a,b,c);
    /* scramble message according to array b */
    bmsg[NLEN]=amsg[NLEN];    /* null terminating character */
    for (i=1;i<=NLEN;i++) {
        j=b[i];
        bmsg[i-1]=amsg[j-1];
    }
    printf("\nscrambled message:\n%s\n",bmsg);
    /* unscramble according to array c */
    cmsg[NLEN]=amsg[NLEN];
    for (i=1;i<=NLEN;i++) {
        j=c[i];
        cmsg[j-1]=bmsg[i-1];
    }
    printf("\nmirrored message:\n%s\n",cmsg);
    free_vector(c,1,NLEN);
    free_vector(b,1,NLEN);
    free_vector(a,1,NLEN);
}
```

rank is a procedure that is similar to indexx. Instead of producing an indexing array, though, it produces a rank table. For an array a[j] and rank table irank[j], entry j in irank will tell what index a[j] will have if a is sorted. irank actually takes its input information not from the array itself, but from the index array produced by indexx. Sample program xrank.c begins with the array from tarray, and feeds it to indexx and rank. The table of ranks produced is listed. To check it, the array a is copied into an array b in the rank order suggested by irank. b should then be in proper order.

```
/* Driver for routine RANK */

#include <stdio.h>
#include "nr.h"
#include "nrutil.h"

#define NP 100
#define MAXSTR 80

main()
{
    char txt[MAXSTR];
    int i,j,k,l,*indx,*irank;
    float *a,b[11];
    FILE *fp;

    indx=ivector(1,NP);
    irank=ivector(1,NP);
    a=vector(1,NP);
    if ((fp = fopen("tarray.dat","r")) == NULL)
        nrerror("Data file TARRAY.DAT not found\n");
    fgets(txt,MAXSTR,fp);
    for (i=1;i<=NP;i++) fscanf(fp,"%f",&a[i]);
    fclose(fp);
    indexx(NP,a,indx);
    rank(NP,indx,irank);
    printf("original array is:\n");
    for (i=0;i<=9;i++) {
        for (j=1;j<=10;j++) printf("%7.2f",a[10*i+j]);
        printf("\n");
    }
    printf("table of ranks is:\n");
    for (i=0;i<=9;i++) {
        for (j=1;j<=10;j++) printf("%7d",irank[10*i+j]);
        printf("\n");
    }
    printf("press return to continue...\n");
    getchar();
    printf("array sorted according to rank table:\n");
    for (i=0;i<=9;i++) {
        for (j=1;j<=10;j++) {
            k=10*i+j;
            for (l=1;l<=NP;l++)
                if (irank[l] == k) b[j]=a[l];
        }
        for (j=1;j<=10;j++) printf("%7.2f",b[j]);
        printf("\n");
    }
    free_vector(a,1,NP);
    free_ivector(irank,1,NP);
    free_ivector(indx,1,NP);
}
```

qcksrt sorts an array by the Quicksort algorithm. Its calling sequence is exactly like that of piksrt and sort, so we again rely on the same sample program, now called xqcksrt.c.

/

```
/* Driver for routine QCKSRT */

#include <stdio.h>
#include "nr.h"
#include "nrutil.h"

#define MAXSTR 80
#define NP 100

main()
{
    char txt[MAXSTR];
    int i,j;
    float *a;
    FILE *fp;

    a=vector(1,NP);
    if ((fp = fopen("tarray.dat","r")) == NULL)
        nrerror("Data file TARRAY.DAT not found\n");
    fgets(txt,MAXSTR,fp);
    for (i=1;i<=NP;i++) fscanf(fp,"%f",&a[i]);
    fclose(fp);
    printf("\noriginal array:\n");
    for (i=0;i<=9;i++) {
        for (j=1;j<=10;j++) printf("%7.2f",a[10*i+j]);
        printf("\n");
    }
    qcksrt(NP,a);
    printf("\nsorted array:\n");
    for (i=0;i<=9;i++) {
        for (j=1;j<=10;j++) printf("%7.2f",a[10*i+j]);
        printf("\n");
    }
    free_vector(a,1,NP);
}
```

Procedure eclass generates a list of equivalence classes for the elements of an input array, based on the arrays lista[j] and listb[j] which list equivalent pairs for each j. In sample program xeclass.c, these lists are

$$\text{lista:} \quad 1,1,5,2,6,2,7,11,3,4,12$$
$$\text{listb:} \quad 5,9,13,6,10,14,3,7,15,8,4$$

According to these lists, 1 is equivalent to 5, 1 is equivalent to 9, etc. If you work it out, you will find the following classes:

$$\text{class1:} \quad 1,5,9,13$$
$$\text{class2:} \quad 2,6,10,14$$
$$\text{class3:} \quad 3,7,11,15$$
$$\text{class4:} \quad 4,8,12$$

The sample program prints out the classes and ought to agree with this list.

```
/* Driver for routine ECLASS */

#include <stdio.h>
#include "nr.h"
#include "nrutil.h"

#define M 11
#define N 15

main()
{
    int i,j,k,lclas,nclass,*nf,*nflag,*nsav;
    static int lista[]={0,1,1,5,2,6,2,7,11,3,4,12},
            listb[]={0,5,9,13,6,10,14,3,7,15,8,4};

    nf=ivector(1,N);
    nflag=ivector(1,N);
    nsav=ivector(1,N);
    eclass(nf,N,lista,listb,M);
    for (i=1;i<=N;i++) nflag[i]=1;
    printf("\nNumbers from 1-%d divided according to\n",N);
    printf("their value modulo 4:\n\n");
    lclas=0;
    for (i=1;i<=N;i++) {
        nclass=nf[i];
        if (nflag[nclass]) {
            nflag[nclass]=0;
            lclas++;
            k=0;
            for (j=i;j<=N;j++)
                if (nf[j] == nf[i]) nsav[++k]=j;
            printf("Class %2d:      ",lclas);
            for (j=1;j<=k;j++) printf("%3d",nsav[j]);
            printf("\n");
        }
    }
    free_ivector(nsav,1,N);
    free_ivector(nflag,1,N);
    free_ivector(nf,1,N);
}
```

eclazz performs the same analysis but figures the equivalences from a boolean function equiv(i,j) that tells whether i and j are in the same equivalence class. In xeclazz.c, equiv is defined as TRUE if (i MOD 4) and (j MOD 4) are the same. It is otherwise FALSE.

```
/* Driver for routine ECLAZZ */

#include <stdio.h>
#include "nr.h"
#include "nrutil.h"

#define N 15

int equiv(i,j)
int i,j;
{
```

```
        return (i % 4) == (j % 4);
}

main()
{
    int i,j,k,lclas,nclass,*nf,*nflag,*nsav;

    nf=ivector(1,N);
    nflag=ivector(1,N);
    nsav=ivector(1,N);
    eclazz(nf,N,equiv);
    for (i=1;i<=N;i++) nflag[i]=1;
    printf("\nNumbers from 1-%d divided according to\n",N);
    printf("their value modulo 4:\n");
    lclas=0;
    for (i=1;i<=N;i++) {
        nclass=nf[i];
        if (nflag[nclass]) {
            nflag[nclass]=0;
            lclas++;
            k=0;
            for (j=i;j<=N;j++)
                if (nf[j] == nf[i]) nsav[++k]=j;
            printf("Class %2d:        ",lclas);
            for (j=1;j<=k;j++) printf("%3d",nsav[j]);
            printf("\n");
        }
    }
    free_ivector(nsav,1,N);
    free_ivector(nflag,1,N);
    free_ivector(nf,1,N);
}
```

Appendix

File `tarray.dat`:

```
Test data for chapter 8:
  29.82 71.51  3.30 87.44 53.42 63.16 89.10 25.75 93.16 27.72
  71.58 48.34 53.11 18.34 27.13 60.31 83.34 22.81 66.84 52.91
  53.42 15.22  8.01 53.39 76.12 79.09 67.61 38.39 24.81 73.21
  13.42 52.10 34.86 99.83 38.46 81.59 61.75 79.62 93.39  3.21
  99.34 92.22 94.29  7.03  6.67 89.35 83.14  9.01 12.68 62.22
   2.95 85.02 95.82 73.96 49.29 77.72 36.65  3.48 48.98 71.83
   1.41  9.48 32.37 89.95 28.39 79.36 54.05 46.08 11.67 37.78
  77.17 74.33 10.13  4.62 49.95 68.40 19.40 34.06  4.11 98.40
  42.44 64.14 89.41 52.99 71.79  3.94 19.73 44.91 71.44 59.10
  27.54 15.67 67.95 55.61 26.05 25.01 82.09 89.67 57.08 38.27
```

Chapter 9: Root Finding and Sets of Equations

Chapter 9 of Numerical Recipes deals primarily with the problem of find-ing roots to equations, and treats the problem in greatest detail in one dimension. We begin with a general-purpose routine called scrsho *that produces a crude graph of a given function on a specified interval. It is used for low-resolution plotting to investigate the properties of the func-tion. With this in hand, we add bracketing routines* zbrac *and* zbrak. *The first of these takes a function and an interval, and expands the interval geometrically until it brackets a root. The second breaks the interval into N subintervals of equal size. It then reports any intervals that contain at least one root. Once bracketed, roots can be found by a number of other routines.* rtbis *finds such roots by bisection.* rtflsp *and* rtsec *use the method of false position and the secant method, respectively.* zbrent *uses a combination of methods to give assured and relatively efficient convergence.* rtnewt *implements the Newton-Raphson root finding method, while* rt-safe *combines it with bisection to correct for its risky global convergence properties.*

For finding the roots of polynomials, laguer *is handy, and when com-bined with its driver* zroots *it can find all roots of a polynomial having complex coefficients. When you have some tentative complex roots of a real polynomial, they can be polished by* qroot, *which employs Bairstow's method.*

In multiple dimensions, root-finding requires some foresight. However, if you can identify the neighborhood of a root of a system of nonlinear equations, then mnewt *will help you to zero in using Newton-Raphson.*

$$\star \quad \star \quad \star \quad \star$$

scrsho is a primitive graphing routine that will print graphs on virtually any terminal or printer. Sample program xscrsho.c demonstrates it by graphing the zero-order Bessel function J_0.

```
/* Driver for routine SCRSHO */

#include <stdio.h>
#include "nr.h"

static float fx(x)
float x;
{
    return bessj0(x);
}
```

```
main()
{
    scrsho(fx);
}
```

zbrac is a root-bracketing routine that works by expanding the range of an interval geometrically until it brackets a root. Sample program xzbrac.c applies it to the Bessel function J_0. It starts with the ten intervals (1.0, 2.0), (2.0, 3.0), etc., and expands each until it contains a root. Then it prints the interval limits, and the function J_0 evaluated at these limits. The two values of J_0 should have opposite signs.

```
/* Driver for routine ZBRAC */

#include <stdio.h>
#include "nr.h"

static float fx(x)
float x;
{
    return bessj0(x);
}

main()
{
    int succes,i;
    float x1,x2;

    printf("%21s %23s\n","bracketing values:","function values:");
    printf("%6s %10s %21s %12s\n","x1","x2","bessj0(x1)","bessj0(x2)");
    for (i=1;i<=10;i++) {
        x1=i;
        x2=x1+1.0;
        succes=zbrac(fx,&x1,&x2);
        if (succes) {
            printf("%7.2f %10.2f %6s %12.6f %12.6f \n",
                x1,x2," ",fx(x1),fx(x2));
        }
    }
}
```

zbrak is much like zbrac except that it takes an interval and subdivides it into N equal parts. It then identifies any of the subintervals that contain roots. Sample program xzbrak.c looks for roots of $J_0(x)$ between x1 = 1.0 and x2 = 50.0 by allowing zbrak to divide the interval into $N = 100$ parts. If there are no roots spaced closer than $\Delta x = 0.49$, then it will find brackets for all roots in this region. The limits of bracketing intervals, as well as function values at these limits, are printed, and again the function values at the end of each interval ought to be of opposite sign. There are 16 roots of J_0 between 1 and 50.

```
/* Driver for routine ZBRAK */

#include <stdio.h>
#include "nr.h"
#include "nrutil.h"

#define N 100
```

```
#define NBMAX 20
#define X1 1.0
#define X2 50.0

static float fx(x)
float x;
{
    return bessj0(x);
}

main()
{
    int i,nb=NBMAX;
    float *xb1,*xb2;

    xb1=vector(1,NBMAX);
    xb2=vector(1,NBMAX);
    zbrak(fx,X1,X2,N,xb1,xb2,&nb);
    printf("\nbrackets for roots of bessj0:\n");
    printf("%21s %10s %16s %10s\n","lower","upper","f(lower)","f(upper)");
    for (i=1;i<=nb;i++)
        printf("%s %2d   %10.4f %10.4f %3s %10.4f %10.4f\n",
            " root ",i,xb1[i],xb2[i]," ",
            fx(xb1[i]),fx(xb2[i]));
    free_vector(xb2,1,NBMAX);
    free_vector(xb1,1,NBMAX);
}
```

Routine `rtbis` begins with the brackets for a root and finds the root itself by bisection, The accuracy with which the root is found is determined by parameter `xacc`. Sample program `xrtbis.c` finds all the roots of Bessel function $J_0(x)$ between $x1 = 1.0$ and $x2 = 50.0$. In this case `xacc` is specified to be about 10^{-6} of the value of the root itself (actually, 10^{-6} of the center of the interval being bisected). The roots are listed, as well as $J_0(\text{root})$ to verify their accuracy.

```
/* Driver for routine RTBIS */

#include <stdio.h>
#include "nr.h"
#include "nrutil.h"

#define N 100
#define NBMAX 20
#define X1 1.0
#define X2 50.0

static float fx(x)
float x;
{
    return bessj0(x);
}

main()
{
    int i,nb=NBMAX;
    float xacc,root,*xb1,*xb2;
```

```
    xb1=vector(1,NBMAX);
    xb2=vector(1,NBMAX);
    zbrak(fx,X1,X2,N,xb1,xb2,&nb);
    printf("\nRoots of bessj0:\n");
    printf("%21s %15s\n","x","f(x)");
    for (i=1;i<=nb;i++) {
        xacc=(1.0e-6)*(xb1[i]+xb2[i])/2.0;
        root=rtbis(fx,xb1[i],xb2[i],xacc);
        printf("root %3d %14.6f %14.6f\n",i,root,fx(root));
    }
    free_vector(xb2,1,NBMAX);
    free_vector(xb1,1,NBMAX);
}
```

The next five sample programs, are essentially identical to the one just discussed, except for the root-finder they employ. `xrtflsp.c` finds the root by "false position". `xrtsec.c` uses the secant method. `xzbrent.c` uses `zbrent` to give reliable and efficient convergence. The Newton-Raphson method implemented in `rtnewt` is demonstrated by `xrtnewt.c`, and `xrtsafe.c` calls `rtsafe`, which improves upon `rtnewt` by combining it with bisection to achieve better global convergence. The latter two programs include a procedure `funcd` that returns the value of the function and its derivative at a given x. In the case of test function $J_0(x)$ the derivative is $-J_1(x)$, and is conveniently in our collection of special functions.

```
/* Driver for routine RTFLSP */

#include <stdio.h>
#include "nr.h"
#include "nrutil.h"

#define N 100
#define NBMAX 20
#define X1 1.0
#define X2 50.0

static float fx(x)
float x;
{
    return bessj0(x);
}

main()
{
    int i,nb=NBMAX;
    float xacc,root,*xb1,*xb2;

    xb1=vector(1,NBMAX);
    xb2=vector(1,NBMAX);
    zbrak(fx,X1,X2,N,xb1,xb2,&nb);
    printf("\nRoots of bessj0:\n");
    printf("%21s %15s\n","x","f(x)");
    for (i=1;i<=nb;i++) {
        xacc=(1.0e-6)*(xb1[i]+xb2[i])/2.0;
        root=rtflsp(fx,xb1[i],xb2[i],xacc);
        printf("root %3d %14.6f %14.6f\n",i,root,fx(root));
    }
```

```
      free_vector(xb2,1,NBMAX);
      free_vector(xb1,1,NBMAX);
}
/* Driver for routine RTSEC */

#include <stdio.h>
#include "nr.h"
#include "nrutil.h"

#define N 100
#define NBMAX 20
#define X1 1.0
#define X2 50.0

static float fx(x)
float x;
{
      return bessj0(x);
}

main()
{
      int i,nb=NBMAX;
      float xacc,root,*xb1,*xb2;

      xb1=vector(1,NBMAX);
      xb2=vector(1,NBMAX);
      zbrak(fx,X1,X2,N,xb1,xb2,&nb);
      printf("\nRoots of bessj0:\n");
      printf("%21s %15s\n","x","f(x)");
      for (i=1;i<=nb;i++) {
           xacc=(1.0e-6)*(xb1[i]+xb2[i])/2.0;
           root=rtsec(fx,xb1[i],xb2[i],xacc);
           printf("root %3d %14.6f %14.6f\n",i,root,fx(root));
      }
      free_vector(xb2,1,NBMAX);
      free_vector(xb1,1,NBMAX);
}
/* Driver for routine ZBRENT */

#include <stdio.h>
#include "nr.h"
#include "nrutil.h"

#define N 100
#define NBMAX 20
#define X1 1.0
#define X2 50.0

static float fx(x)
float x;
{
      return bessj0(x);
}

main()
```

```
{
    int i,nb=NBMAX;
    float tol,root,*xb1,*xb2;

    xb1=vector(1,NBMAX);
    xb2=vector(1,NBMAX);
    zbrak(fx,X1,X2,N,xb1,xb2,&nb);
    printf("\nRoots of bessj0:\n");
    printf("%21s %15s\n","x","f(x)");
    for (i=1;i<=nb;i++) {
        tol=(1.0e-6)*(xb1[i]+xb2[i])/2.0;
        root=zbrent(fx,xb1[i],xb2[i],tol);
        printf("root %3d %14.6f %14.6f\n",i,root,fx(root));
    }
    free_vector(xb2,1,NBMAX);
    free_vector(xb1,1,NBMAX);
}

/* Driver for routine RTNEWT */

#include <stdio.h>
#include "nr.h"
#include "nrutil.h"

#define N 100
#define NBMAX 20
#define X1 1.0
#define X2 50.0

static float fx(x)
float x;
{
    return bessj0(x);
}

static void funcd(x,fn,df)
float x,*fn,*df;
{
    *fn=bessj0(x);
    *df = -bessj1(x);
}

main()
{
    int i,nb=NBMAX;
    float xacc,root,*xb1,*xb2;

    xb1=vector(1,NBMAX);
    xb2=vector(1,NBMAX);
    zbrak(fx,X1,X2,N,xb1,xb2,&nb);
    printf("\nRoots of bessj0:\n");
    printf("%21s %15s\n","x","f(x)");
    for (i=1;i<=nb;i++) {
        xacc=(1.0e-6)*(xb1[i]+xb2[i])/2.0;
        root=rtnewt(funcd,xb1[i],xb2[i],xacc);
        printf("root %3d %14.6f %14.6f\n",i,root,fx(root));
    }
```

```
        free_vector(xb2,1,NBMAX);
        free_vector(xb1,1,NBMAX);
}

/* Driver for routine RTSAFE */

#include <stdio.h>
#include "nr.h"
#include "nrutil.h"

#define N 100
#define NBMAX 20
#define X1 1.0
#define X2 50.0

static float fx(x)
float x;
{
    return bessj0(x);
}

static void funcd(x,fn,df)
float x,*fn,*df;
{
    *fn=bessj0(x);
    *df = -bessj1(x);
}

main()
{
    int i,nb=NBMAX;
    float xacc,root,*xb1,*xb2;

    xb1=vector(1,NBMAX);
    xb2=vector(1,NBMAX);
    zbrak(fx,X1,X2,N,xb1,xb2,&nb);
    printf("\nRoots of bessj0:\n");
    printf("%21s %15s\n","x","f(x)");
    for (i=1;i<=nb;i++) {
        xacc=(1.0e-6)*(xb1[i]+xb2[i])/2.0;
        root=rtsafe(funcd,xb1[i],xb2[i],xacc);
        printf("root %3d %14.6f %14.6f\n",i,root,fx(root));
    }
    free_vector(xb2,1,NBMAX);
    free_vector(xb1,1,NBMAX);
}
```

Routine `laguer` finds the roots of a polynomial with complex coefficients. The polynomial of degree M is specified by $M + 1$ coefficients which, in sample program `xlaguer.c`, are specified in the complex array a of dimension $M + 1$. The polynomial in this case is

$$F(x) = x^4 - (1 + 2i)x^2 + 2i$$

The four roots of this polynomial are $x = 1.0$, $x = -1.0$, $x = 1 + i$, and $x = -(1 + i)$. `laguer` proceeds on the basis of a trial root, and attempts to converge to true roots. The root to which it converges depends on the trial value. The program tries a series of complex trial values along the line in the imaginary plane from $-1.0 - i$ to $1.0 + i$.

The actual roots to which it converges are compared to all previously found values, and if different, are printed.

```
/* Driver for routine LAGUER */

#include <stdio.h>
#include <math.h>
#include "nr.h"
#include "complex.h"

#define M 4          /* degree of polynomial */
#define MP1 (M+1)    /* no. of polynomial coefficients */
#define NTRY 21
#define NTRY1 NTRY+1
#define EPS 1.e-6

main()
{
    fcomplex y[NTRY1],x;
    static fcomplex a[MP1] = {{0.0,2.0},
                    {0.0,0.0},
                    {-1.0,-2.0},
                    {0.0,0.0},
                    {1.0,0.0} };
    int i,iflag,j,n=0,polish=0;

    printf("\nRoots of polynomial x^4-(1+2i)*x^2+2i\n");
    printf("\n%15s %15s\n","Real","Complex");
    for (i=1;i<=NTRY;i++) {
        x=Complex((i-11.0)/10.0,(i-11.0)/10.0);
        laguer(a,M,&x,EPS,polish);
        if (n == 0) {
            n=1;
            y[1]=x;
            printf("%5d %12.6f %12.6f\n",n,x.r,x.i);
        } else {
            iflag=0;
            for (j=1;j<=n;j++)
                if (Cabs(Csub(x,y[j])) <= EPS*Cabs(x)) iflag=1;
            if (iflag == 0) {
                y[++n]=x;
                printf("%5d %12.6f %12.6f\n",n,x.r,x.i);
            }
        }
    }
}
```

zroots is a driver for laguer. Sample program xzroots.c exercises zroots, using the same polynomial as the previous routine. First it finds the four roots. Then it corrupts each one by multiplying by 1.01. Finally it uses zroots again to polish the corrupted roots by setting the boolean variable polish to TRUE.

```
/* Driver for routine ZROOTS */

#include <stdio.h>
#include "nr.h"
#include "complex.h"
```

```
#define M 4
#define MP1 (M+1)
#define TRUE 1
#define FALSE 0

main()
{
    int i,polish;
    fcomplex roots[MP1];
    static fcomplex a[MP1] = {{0.0,2.0},
                    {0.0,0.0},
                    {-1.0,-2.0},
                    {0.0,0.0},
                    {1.0,0.0} };

    printf("\nRoots of the polynomial x^4-(1+2i)*x^2+2i\n");
    polish=FALSE;
    zroots(a,M,roots,polish);
    printf("\nUnpolished roots:\n");
    printf("%14s %13s %13s\n","root #","real","imag.");
    for (i=1;i<=M;i++)
        printf("%11d %18.6f %12.6f\n",i,roots[i].r,roots[i].i);
    printf("\nCorrupted roots:\n");
    for (i=1;i<=M;i++)
        roots[i]=RCmul(1+0.01*i,roots[i]);
    printf("%14s %13s %13s\n","root #","real","imag.");
    for (i=1;i<=M;i++)
        printf("%11d %18.6f %12.6f\n",i,roots[i].r,roots[i].i);
    polish=TRUE;
    zroots(a,M,roots,polish);
    printf("\nPolished roots:\n");
    printf("%14s %13s %13s\n","root #","real","imag.");
    for (i=1;i<=M;i++)
        printf("%11d %18.6f %12.6f \n",i,roots[i].r,roots[i].i);
}
```

qroot is used for finding quadratic factors of polynomials with real coefficients. In the case of sample program xqroot.c, the polynomial is

$$P(x) = x^6 - 6x^5 + 16x^4 - 24x^3 + 25x^2 - 18x + 10.$$

The program proceeds like that of laguer. Successive trial values for quadratic factors $x^2 + Bx + C$ (in the form of guesses for B and C) are made, and for each trial, qroot converges on correct values. If the B and C which are found are unlike any previous values, then they are printed. By this means, all three quadratic factors are located. You can, of course, compare their product to the polynomial above.

```
/* Driver for routine QROOT */

#include <stdio.h>
#include <math.h>
#include "nr.h"
#include "nrutil.h"

#define N 6      /* degree of polynomial */
#define EPS 1.0e-6
#define NTRY 10
```

```
#define TINY 1.0e-5

main()
{
    int i,j,nflag,nroot;
    static float p[N+1]={10.0,-18.0,25.0,-24.0,16.0,-6.0,1.0};
    float *b,*c;

    b=vector(1,NTRY);
    c=vector(1,NTRY);
    printf("\nP(x)=x^6-6x^5+16x^4-24x^3+25x^2-18x+10\n");
    printf("Quadratic factors x^2+bx+c\n\n");
    printf("%6s %10s %12s \n\n","factor","b","c");
    nroot=0;
    for (i=1;i<=NTRY;i++) {
        c[i]=0.5*i;
        b[i] = -0.5*i;
        qroot(p,N,&b[i],&c[i],EPS);
        if (nroot == 0) {
            printf("%4d %15.6f %12.6f\n",nroot,b[i],c[i]);
            nroot=1;
        } else {
            nflag=0;
            for (j=1;j<=nroot;j++)
                if ((fabs(b[i]-b[j]) < TINY)
                    && (fabs(c[i]-c[j]) < TINY))
                        nflag=1;
            if (nflag == 0) {
                printf("%4d %15.6f %12.6f\n",nroot,b[i],c[i]);
                ++nroot;
            }
        }
    }
    free_vector(c,1,NTRY);
    free_vector(b,1,NTRY);
}
```

Finally, mnewt looks for roots of multiple nonlinear equations. In order to run a sample program xmnewt.c we supply a procedure usrfun that returns the matrix alpha of partial derivatives of the functions with respect to each of the variables, and vector beta, containing the negatives of the function values. The sample program tries to find sets of variables that solve the four equations

$$-x_1^2 - x_2^2 - x_3^2 + x_4 = 0$$
$$x_1^2 + x_2^2 + x_3^2 + x_4^2 - 1 = 0$$
$$x_1 - x_2 = 0$$
$$x_2 - x_3 = 0$$

You will probably be able to find the two solutions to this set even without mnewt, noting that $x_1 = x_2$ and $x_2 = x_3$. If not, simply take the output from mnewt and plug it into these equations for verification. The output from mnewt should convince you of the need for good starting values.

```
/* Driver for routine MNEWT */

#include <stdio.h>
#include <math.h>
#include "nr.h"
#include "nrutil.h"

#define SQR(a) ((a)*(a))

void usrfun(x,alpha,bet)
float *x,**alpha,*bet;
{
    alpha[1][1] = -2.0*x[1];
    alpha[1][2] = -2.0*x[2];
    alpha[1][3] = -2.0*x[3];
    alpha[1][4]=1.0;
    alpha[2][1]=2.0*x[1];
    alpha[2][2]=2.0*x[2];
    alpha[2][3]=2.0*x[3];
    alpha[2][4]=2.0*x[4];
    alpha[3][1]=1.0;
    alpha[3][2] = -1.0;
    alpha[3][3]=0.0;
    alpha[3][4]=0.0;
    alpha[4][1]=0.0;
    alpha[4][2]=1.0;
    alpha[4][3] = -1.0;
    alpha[4][4]=0.0;
    bet[1]=SQR(x[1])+SQR(x[2])+SQR(x[3])-x[4];
    bet[2] = -SQR(x[1])-SQR(x[2])-SQR(x[3])-SQR(x[4])+1.0;
    bet[3] = -x[1]+x[2];
    bet[4] = -x[2]+x[3];
}

#define NTRIAL 5
#define TOLX 1.0e-6
#define N 4
#define TOLF 1.0e-6

main()
{
    int i,j,k,kk;
    float xx,*x,*bet,**alpha;

    alpha=matrix(1,N,1,N);
    bet=vector(1,N);
    x=vector(1,N);
    for (kk=1;kk<=2;kk++) {
        for (k=1;k<=3;k++) {
            xx=0.21*k*(2*kk-3);
            printf("Starting vector number %2d\n",k);
            for (i=1;i<=4;i++) {
                x[i]=xx+0.2*i;
                printf("%7s%1d%s %5.2f\n",
                    "x[",i,"] = ",x[i]);
            }
            printf("\n");
```

```
            for (j=1;j<=NTRIAL;j++) {
                mnewt(1,x,N,TOLX,TOLF);
                usrfun(x,alpha,bet);
                printf("%5s %13s %13s\n","i","x[i]","f");
                for (i=1;i<=N;i++)
                    printf("%5d %14.6f %15.6f\n",
                        i,x[i],-bet[i]);
                printf("\npress RETURN to continue...\n");
                getchar();
            }
        }
    }
    free_vector(x,1,N);
    free_vector(bet,1,N);
    free_matrix(alpha,1,N,1,N);
}
```

Chapter 10: Minimization and Maximization of Functions

Chapter 10 of Numerical Recipes deals with finding the maxima and minima of functions. The task has two parts, first the discovery of one or more bracketing intervals, and then the convergence to an extremum. mnbrak *begins with two specified abscissas of a function and searches in the "downhill" direction for brackets of a minimum.* golden *can then take a bracketing triplet and perform a golden section search to a specified precision, for the minimum itself. When you are not concerned with worst-case examples, but only very efficient average-case performance, Brent's method (routine* brent*) is recommended. In the event that means are at hand for calculating the function's derivative as well as its value, consider* dbrent.

Multidimensional minimization strategies may be based on the one-dimensional algorithms. Our single example of an algorithm that is not *so based is* amoeba*, which utilizes the downhill simplex method. Among the ones that* do *use one-dimensional methods are* powell, frprmn, *and* dfpmin. *These three all make calls to* linmin, *a routine that minimizes a mathematical function along a given direction in space.* linmin *in turn uses the one-dimensional algorithm* brent, *if derivatives are not known, or* dbrent *if they are.* powell *uses only function values and minimizes along an artfully chosen set of favorable directions.* frprmn *uses a Fletcher-Reeves-Polak-Ribiere minimization and requires the calculation of derivatives for the function.* dfpmin *uses a variant of the Davidon-Fletcher-Powell variable metric method. This, too, requires calculation of derivatives.*

The chapter ends with two topics of somewhat different nature. The first is linear programming, which deals with the maximation of a linear combination of variables, subject to linear constraints. This problem is dealt with by the simplex method in routine simplx. *The second is the subject of large scale optimization, which is illustrated with the method of simulated annealing, and applied particularly to the "travelling salesman" problem in routine* anneal.

★ ★ ★ ★

mnbrak searches a given function for a minimum. Given two values ax and bx of abscissa, it searches in the downward direction until it can find three new values ax, bx, cx that bracket a minimum. fa, fb, fc are the values of the function at these points. Sample program xmnbrak.c is a simple application of mnbrak applied to the

Bessel function J_0. It tries a series of starting values `ax,bx` each encompassing an interval of length 1.0. `mnbrak` then finds several bracketing intervals of various minima of J_0.

```
/* Driver for routine MNBRAK */

#include <stdio.h>
#include "nr.h"

float func(x)
float x;
{
    return bessj0(x);
}

main()
{
    float ax,bx,cx,fa,fb,fc;
    int i;

    for (i=1;i<=10;i++) {
        ax=i*0.5;
        bx=(i+1.0)*0.5;
        mnbrak(&ax,&bx,&cx,&fa,&fb,&fc,func);
        printf("%14s %12s %12s\n","a","b","c");
        printf("%3s %14.6f %12.6f %12.6f\n","x",ax,bx,cx);
        printf("%3s %14.6f %12.6f %12.6f\n","f",fa,fb,fc);
    }
}
```

(handwritten note: void)

Routine `golden` continues the minimization process by taking a bracketing triplet `ax,bx,cx` and performing a golden section search to isolate the contained minimum to a stated precision `TOL`. Sample program `xgolden.c` again uses J_0 as the test function. Using intervals `(ax,bx)` of length 1.0 it uses `mnbrak` to bracket all minima between $x = 0.0$ and $x = 100.0$. Some minima are bracketed more than once. On each pass, the bracketed solution is tracked down by `golden`. It is then compared to all previously located minima, and if different it is added to the collection by incrementing `nmin` (number of minima found) and adding the location `xmin` of the minima to the list in array `amin`. As a check of `golden`, the routine prints out the value of J_0 at the minimum, and also the value of J_1, which ought to be zero at extrema of J_0.

```
/* Driver for routine GOLDEN */

#include <stdio.h>
#include <math.h>
#include "nr.h"

#define TOL 1.0e-6
#define EQL 1.0e-3

float func(x)
float x;
{
    return bessj0(x);
}
```

```
main()
{
    int i,iflag,j,nmin=0;
    float ax,bx,cx,fa,fb,fc,xmin,gold,amin[21];

    printf("Minima of the function bessj0\n");
    printf("%10s %8s %17s %12s\n","min. #","x","bessj0(x)","bessj1(x)");
    for (i=1;i<=100;i++) {
        ax=i;
        bx=i+1.0;
        mnbrak(&ax,&bx,&cx,&fa,&fb,&fc,func);
        gold=golden(ax,bx,cx,func,TOL,&xmin);
        if (nmin == 0) {
            amin[1]=xmin;
            nmin=1;
            printf("%7d %15.6f %12.6f %12.6f\n",
                nmin,xmin,bessj0(xmin),bessj1(xmin));
        } else {
            iflag=0;
            for (j=1;j<=nmin;j++)
                if (fabs(xmin-amin[j]) <= EQL*xmin) iflag=1;
            if (iflag == 0) {
                amin[++nmin]=xmin;
                printf("%7d %15.6f %12.6f %12.6f\n",
                    nmin,xmin,bessj0(xmin),bessj1(xmin));
            }
        }
    }
}
```

There are two other routines presented which also take the bracketing triplet ax, bx, cx from mnbrak and find the contained minimum. They are brent and dbrent. The sample programs for these two, xbrent.c and xdbrent.c, are virtually identical to that used on golden. Note that dbrent is only used when the derivative, in this case (−bessj1), can be calculated conveniently.

```
/* Driver for routine BRENT */

#include <stdio.h>
#include <math.h>
#include "nr.h"

#define TOL 1.0e-6
#define EQL 1.0e-4

float func(x)
float x;
{
    return bessj0(x);
}

main()
{
    int i,iflag,j,nmin=0;
    float ax,bx,cx,fa,fb,fc,xmin,bren,amin[21];

    printf("\nMinima of the function bessj0\n");
```

```
        ("%10s %8s %17s %12s\n","min. #","x","bessj0(x)","bessj1(x)");
     i=1;i<=100;i++) {
    ax=i;
    bx=i+1.0;
    mnbrak(&ax,&bx,&cx,&fa,&fb,&fc,func);
    bren=brent(ax,bx,cx,func,TOL,&xmin);
    if (nmin == 0) {
        amin[1]=xmin;
        nmin=1;
        printf("%7d %15.6f %12.6f %12.6f\n",
            nmin,xmin,bessj0(xmin),bessj1(xmin));
    } else {
        iflag=0;
        for (j=1;j<=nmin;j++)
            if (fabs(xmin-amin[j]) <= (EQL*xmin)) iflag=1;
        if (iflag == 0) {
            amin[++nmin]=xmin;
            printf("%7d %15.6f %12.6f %12.6f\n",
                nmin,xmin,bessj0(xmin),bessj1(xmin));
        }
    }
  }
}

/* Driver for routine DBRENT */

#include <stdio.h>
#include <math.h>
#include "nr.h"

#define TOL 1.0e-6
#define EQL 1.0e-4

float dfunc(x)
float x;
{
    return -bessj1(x);
}

float func(x)
float x;
{
    return bessj0(x);
}

main()
{
    int i,iflag,j,nmin=0;
    float ax,bx,cx,fa,fb,fc,xmin,dbr,amin[21];

    printf("\nMinima of the function bessj0\n");
    printf("%10s %8s %16s %12s %11s\n",
        "min. #","x","bessj0(x)","bessj1(x)","DBRENT");
    for (i=1;i<=100;i++) {
        ax=i;
        bx=i+1.0;
        mnbrak(&ax,&bx,&cx,&fa,&fb,&fc,func);
```

```
    dbr=dbrent(ax,bx,cx,func,dfunc,TOL,&xmin);
    if (nmin == 0) {
        amin[1]=xmin;
        nmin=1;
        printf("%7d %15.6f %12.6f %12.6f %12.6f\n",
            nmin,xmin,func(xmin),dfunc(xmin),dbr);
    } else {
        iflag=0;
        for (j=1;j<=nmin;j++)
            if (fabs(xmin-amin[j]) <= EQL*xmin) iflag=1;
        if (iflag == 0) {
            amin[++nmin]=xmin;
            printf("%7d %15.6f %12.6f %12.6f %12.6f\n",
                nmin,xmin,func(xmin),dfunc(xmin),dbr);
        }
    }
  }
}
```

Numerical Recipes presents several methods for minimization in multiple dimensions. Among these, the downhill simplex method carried out by amoeba was the only one that did not treat the problem as a series of one-dimensional minimizations. As input, amoeba requires the coordinates of $N+1$ vertices of a starting simplex in N-dimensional space, and the values y of the function at each of these vertices. Sample program xamoeba.c tries the method out on the exotic function

$$\text{func} = 0.6 - J_0[(x-0.5)^2 + (y-0.6)^2 + (z-0.7)^2]$$

which has a minimum at $(x, y, z) = (0.5, 0.6, 0.7)$. As vertices of the starting simplex, specified by the array p, we used $(0, 0, 0)$, $(1, 0, 0)$, $(0, 1, 0)$, and $(0, 0, 1)$. A vector x[i] is set successively to each vertex to allow the evaluation of function values y. This data is submitted to amoeba along with FTOL=1.0e-6 to specify the tolerance on the function value. The vertices and corresponding function values of the final simplex are printed out, and you can easily check whether the specified tolerance is met.

```
/* Driver for routine AMOEBA */

#include <stdio.h>
#include <math.h>
#include "nr.h"
#include "nrutil.h"

#define MP 4
#define NP 3
#define FTOL 1.0e-6
#define SQR(a) ((a)*(a))

float func(x)
float x[];
{
    return 0.6-bessj0(SQR(x[1]-0.5)+SQR(x[2]-0.6)+SQR(x[3]-0.7));
}

main()
{
    int i,nfunc,j,ndim=NP;
```

```
        float *x,*y,**p;

        x=vector(1,NP);
        y=vector(1,MP);
        p=matrix(1,MP,1,NP);
        for (i=1;i<=MP;i++) {
            for (j=1;j<=NP;j++)
                x[j]=p[i][j]=(i == (j+1) ? 1.0 : 0.0);
            y[i]=func(x);
        }
        amoeba(p,y,ndim,FTOL,func,&nfunc);
        printf("\nNo. of function evaluations: %3d\n",nfunc);
        printf("Vertices of final 3-d simplex and\n");
        printf("function values at the vertices:\n\n");
        printf("%3s %10s %12s %12s %14s\n\n",
            "i","x[i]","y[i]","z[i]","function");
        for (i=1;i<=MP;i++) {
            printf("%3d ",i);
            for (j=1;j<=NP;j++) printf("%12.6f ",p[i][j]);
            printf("%12.6f\n",y[i]);
        }
        printf("\nTrue minimum is at (0.5,0.6,0.7)\n");
        free_matrix(p,1,MP,1,NP);
        free_vector(y,1,MP);
        free_vector(x,1,NP);
}
```

`powell` carries out one-dimensional minimizations along favorable directions in N-dimensional space. The function minimized in sample program `xpowell.c` is defined as

$$\text{func}(x,y,z) = \tfrac{1}{2} - J_0[(x-1)^2 + (y-2)^2 + (z-3)^2].$$

The program provides `powell` with a starting point P of $(3/2, 3/2, 5/2)$ and a set of initial directions, here chosen to be the unit directions $(1,0,0)$, $(0,1,0)$, and $(0,0,1)$. `powell` performs its one-dimensional minimizations with `linmin`, which is discussed next.

```
/* Driver for routine POWELL */

#include <stdio.h>
#include <math.h>
#include "nr.h"
#include "nrutil.h"

#define NDIM 3
#define FTOL 1.0e-6
#define SQR(a) ((a)*(a))

float func(x)
float x[];
{
    return 0.5-bessj0(SQR(x[1]-1.0)+SQR(x[2]-2.0)+SQR(x[3]-3.0));
}

main()
{
    int i,iter,j;
```

```
       float fret,**xi;
       static float p[]={0.0,1.5,1.5,2.5};

       xi=matrix(1,NDIM,1,NDIM);
       for (i=1;i<=NDIM;i++)
           for (j=1;j<=NDIM;j++)
               xi[i][j]=(i == j ? 1.0 : 0.0);
       powell(p,xi,NDIM,FTOL,&iter,&fret,func);
       printf("Iterations: %3d\n\n",iter);
       printf("Minimum found at: \n");
       for (i=1;i<=NDIM;i++) printf("%12.6f",p[i]);
       printf("\n\nMinimum function value = %12.6f \n\n",fret);
       printf("True minimum of function is at:\n");
       printf("%12.6f %12.6f %12.6f\n",1.0,2.0,3.0);
       free_matrix(xi,1,NDIM,1,NDIM);
}
```

linmin, as we have said, finds the minimum of a function along a direction in N-dimensional space. To use it we specify a point P and a direction vector xi, both in N-space. linmin then does the book-keeping required to treat the function as a function of position along this line, and minimizes the function with a conventional one-dimensional minimization routine. Sample program xlinmin.c feeds linmin the function

$$\text{func}(x,y,z) = (x-1)^2 + (y-1)^2 + (z-1)^2$$

which has a minimum at $(x,y,z) = (1,1,1)$. It also chooses point P to be the origin $(0,0,0)$, and tries a series of directions

$$\left(\sqrt{2}\cos\left(\frac{\pi}{2}\frac{i}{10.0}\right), \quad \sqrt{2}\sin\left(\frac{\pi}{2}\frac{i}{10.0}\right), \quad 1.0 \right) \qquad i = 1,\ldots,10$$

For each pass, the location of the minimum, and the value of the function at the minimum, are printed. Among the directions searched is the direction $(1,1,1)$. Along this direction, of course, the minimum function value should be zero and should occur at $(1,1,1)$.

```
/* Driver for routine LINMIN */

#include <stdio.h>
#include <math.h>
#include "nr.h"
#include "nrutil.h"

#define NDIM 3
#define PIO2 1.5707963

float func(x)
float x[];
{
    int i;
    float f=0.0;

    for (i=1;i<=3;i++) f += (x[i]-1.0)*(x[i]-1.0);
    return f;
}

main()
{
```

```
      int i,j;
      float fret,sr2,x,*p,*xi;

      p=vector(1,NDIM);
      xi=vector(1,NDIM);
      printf("\nMinimum of a 3-d quadratic centered\n");
      printf("at (1.0,1.0,1.0). Minimum is found\n");
      printf("along a series of radials.\n\n");
      printf("%9s %12s %12s %14s \n","x","y","z","minimum");
      for (i=0;i<=10;i++) {
          x=PIO2*i/10.0;
          sr2=sqrt(2.0);
          xi[1]=sr2*cos(x);
          xi[2]=sr2*sin(x);
          xi[3]=1.0;
          p[1]=p[2]=p[3]=0.0;
          linmin(p,xi,NDIM,&fret,func);
          for (j=1;j<=3;j++) printf("%12.6f ",p[j]);
          printf("%12.6f\n",fret);
      }
      free_vector(xi,1,NDIM);
      free_vector(p,1,NDIM);
}
```

f1dim accompanies linmin and is the routine that makes an N-dimensional function effectively a one-dimensional function along a given line in N-space. There is little to check here, and our perfunctory demonstration of its use, in sample program xf1dim.c, simply plots f1dim as a one-dimensional function, given the function

$$\text{func}(x,y,z) = (x-1)^2 + (y-1)^2 + (z-1)^2.$$

You get to choose the direction; then scrsho plots the function along this direction. Try the direction $(1,1,1)$ along which you should find a minimum value of func=0 at position $(1,1,1)$.

```
/* Driver for routine F1DIM */

#include <stdio.h>
#include "nr.h"
#include "nrutil.h"

float func(x)
float x[];
{
    int i;
    float f=0.0;

    for (i=1;i<=3;i++) f += (x[i]-1.0)*(x[i]-1.0);
    return f;
}

#define NDIM 3

int ncom=0;      /* defining declarations */
float *pcom=0,*xicom=0,(*nrfunc)();

main()
```

```
{
    ncom=NDIM;
    pcom=vector(1,ncom);
    xicom=vector(1,ncom);
    nrfunc=func;
    pcom[1]=pcom[2]=pcom[3]=0.0;
    printf("\nEnter vector direction along which to\n");
    printf("plot the function. Minimum is in the\n");
    printf("direction 1.0 1.0 1.0 - enter x y z:\n");
    scanf(" %f %f %f",&xicom[1],&xicom[2],&xicom[3]);
    scrsho(f1dim);
    free_vector(xicom,1,ncom);
    free_vector(pcom,1,ncom);
}
```

frprmn is another multidimensional minimizer that relies on the one-dimensional minimizations of linmin. It works, however, via the Fletcher-Reeves-Polak-Ribiere method and requires that routines be supplied for calculating both the function and its gradient. Sample program xfrprmn.c, for example, uses

$$\text{func}(x,y,z) = 1.0 - J_0(x - \tfrac{1}{2})J_0(y - \tfrac{1}{2})J_0(z - \tfrac{1}{2})$$

and

$$\frac{\partial \text{func}}{\partial x} = J_1(x - \tfrac{1}{2})J_0(y - \tfrac{1}{2})J_0(z - \tfrac{1}{2})$$

etc. A number of trial starting vectors are used, and each time, frprmn manages to find the minimum at $(1/2, 1/2, 1/2)$.

```
/* Driver for routine FRPRMN */

#include <stdio.h>
#include <math.h>
#include "nr.h"
#include "nrutil.h"

#define NDIM 3
#define FTOL 1.0e-6
#define PIO2 1.5707963

float func(x)
float x[];
{
    return 1.0-bessj0(x[1]-0.5)*bessj0(x[2]-0.5)*bessj0(x[3]-0.5);
}

void dfunc(x,df)
float x[],df[];
{
    df[1]=bessj1(x[1]-0.5)*bessj0(x[2]-0.5)*bessj0(x[3]-0.5);
    df[2]=bessj0(x[1]-0.5)*bessj1(x[2]-0.5)*bessj0(x[3]-0.5);
    df[3]=bessj0(x[1]-0.5)*bessj0(x[2]-0.5)*bessj1(x[3]-0.5);
}

main()
{
    int iter,k;
    float angl,fret,*p;
```

```
    p=vector(1,NDIM);
    printf("Program finds the minimum of a function\n");
    printf("with different trial starting vectors.\n");
    printf("True minimum is (0.5,0.5,0.5)\n");
    for (k=0;k<=4;k++) {
        angl=PIO2*k/4.0;
        p[1]=2.0*cos(angl);
        p[2]=2.0*sin(angl);
        p[3]=0.0;
        printf("\nStarting vector: (%6.4f,%6.4f,%6.4f)\n",
            p[1],p[2],p[3]);
        frprmn(p,NDIM,FTOL,&iter,&fret,func,dfunc);
        printf("Iterations: %3d\n",iter);
        printf("Solution vector: (%6.4f,%6.4f,%6.4f)\n",
            p[1],p[2],p[3]);
        printf("Func. value at solution %14f\n",fret);
    }
    free_vector(p,1,NDIM);
}
```

The routine dlinmin is an alternative to linmin for minimization procedures that use derivative information as well as function values. The sample program for dlinmin is essentially the same as for linmin, given above.

```
/* Driver for routine DLINMIN */

#include <stdio.h>
#include <math.h>
#include "nr.h"
#include "nrutil.h"

#define NDIM 3
#define PIO2 1.5707963

float func(x)
float x[];
{
    int i;
    float f=0.0;

    for (i=1;i<=3;i++) f += (x[i]-1.0)*(x[i]-1.0);
    return f;
}

void dfunc(x,df)
float x[],df[];
{
    int i;

    for (i=1;i<=3;i++) df[i]=2.0*(x[i]-1.0);
}

main()
{
    int i,j;
    float fret,sr2,x,*p,*xi;
```

```
        p=vector(1,NDIM);
        xi=vector(1,NDIM);
        printf("\nMinimum of a 3-d quadratic centered\n");
        printf("at (1.0,1.0,1.0). Minimum is found\n");
        printf("along a series of radials.\n\n");
        printf("%9s %12s %12s %14s \n","x","y","z","minimum");
        for (i=0;i<=10;i++) {
            x=PIO2*i/10.0;
            sr2=sqrt(2.0);
            xi[1]=sr2*cos(x);
            xi[2]=sr2*sin(x);
            xi[3]=1.0;
            p[1]=p[2]=p[3]=0.0;
            dlinmin(p,xi,NDIM,&fret,func,dfunc);
            for (j=1;j<=3;j++) printf("%12.6f ",p[j]);
            printf("%12.6f\n",fret);
        }
        free_vector(xi,1,NDIM);
        free_vector(p,1,NDIM);
}
```

Completeness requires that we provide a sample program for df1dim, which is presented in *Numerical Recipes* as a routine for converting the *N*-dimensional gradient procedure to one that provides the first derivative of the function along a specified line in *N*-dimensional space. It is exactly analogous to f1dim and the program xdf1dim.c is the same.

```
/* Driver for routine DF1DIM */

#include <stdio.h>
#include "nr.h"
#include "nrutil.h"

#define NDIM 3

int ncom=0;     /* defining declarations */
float *pcom=0,*xicom=0;
void (*nrdfun)();

void dfunc(x,df)
float x[],df[];
{
    int i;

    for (i=1;i<=3;i++) df[i]=(x[i]-1.0)*(x[i]-1.0);
}

main()
{
    ncom=NDIM;
    pcom=vector(1,ncom);
    xicom=vector(1,ncom);
    nrdfun=dfunc;
    printf("\nEnter vector direction along which to\n");
    printf("plot the function. Minimum is in the\n");
    printf("direction 1.0 1.0 1.0 - enter x y z:\n\n");
```

```
        pcom[1]=pcom[2]=pcom[3]=0.0;
        scanf("%f %f %f",&xicom[1],&xicom[2],&xicom[3]);
        scrsho(df1dim);
        free_vector(xicom,1,ncom);
        free_vector(pcom,1,ncom);
}
```

dfpmin implements the Broyden-Fletcher-Goldfarb-Shanno variant of the Davidon-Fletcher-Powell minimization by variable metric methods. It requires somewhat more intermediate storage than the preceding routine and is not considered superior in other ways. However, it is a popular method. Sample program xdfpmin.c works just as did the program for frprmn, including the fact that it requires a procedure for calculation of the derivative.

```
/* Driver for routine DFPMIN */

#include <stdio.h>
#include <math.h>
#include "nr.h"
#include "nrutil.h"

#define NDIM 3
#define FTOL 1.0e-6
#define PIO2 1.5707963

float func(x)
float x[];
{
        return 1.0-bessj0(x[1]-0.5)*bessj0(x[2]-0.5)*bessj0(x[3]-0.5);
}

void dfunc(x,df)
float x[],df[];
{
        df[1]=bessj1(x[1]-0.5)*bessj0(x[2]-0.5)*bessj0(x[3]-0.5);
        df[2]=bessj0(x[1]-0.5)*bessj1(x[2]-0.5)*bessj0(x[3]-0.5);
        df[3]=bessj0(x[1]-0.5)*bessj0(x[2]-0.5)*bessj1(x[3]-0.5);
}

main()
{
        int iter,k;
        float angl,fret,*p;

        p=vector(1,NDIM);
        printf("Program finds the minimum of a function\n");
        printf("with different trial starting vectors.\n");
        printf("True minimum is (0.5,0.5,0.5)\n");
        for (k=0;k<=4;k++) {
                angl=PIO2*k/4.0;
                p[1]=2.0*cos(angl);
                p[2]=2.0*sin(angl);
                p[3]=0.0;
                printf("\nStarting vector: (%6.4f,%6.4f,%6.4f)\n",
                        p[1],p[2],p[3]);
                dfpmin(p,NDIM,FTOL,&iter,&fret,func,dfunc);
                printf("Iterations: %3d\n",iter);
```

```
        printf("Solution vector:  (%6.4f,%6.4f,%6.4f)\n",
            p[1],p[2],p[3]);
        printf("Func. value at solution %14f\n",fret);
    }
    free_vector(p,1,NDIM);
}
```

simplx is a procedure for dealing with problems in linear programming. In these problems the goal is to maximize a linear combination of N variables, subject to the constraint that none be negative, and that as a group they satisfy a number of other constraints. In order to clarify the subject, *Numerical Recipes* presents a sample problem in equations (10.8.6) and (10.8.7), translating the problem into tableau format in (10.8.18), and presenting a solution in equation (10.8.19). Sample program xsimplx.c carries out the analysis that leads to this solution.

```c
/* Driver for routine SIMPLX*/
/* Incorporates example discussed in text */

#include <stdio.h>
#include "nr.h"
#include "nrutil.h"

#define N 4
#define M 4
#define NP 5        /* NP >= N+1 */
#define MP 6        /* MP >= M+2 */
#define M1 2        /* M1+M2+M3 = M */
#define M2 1
#define M3 1
#define NM1M2 (N+M1+M2)

main()
{
    int i,icase,j,*izrov,*iposv;
    static float c[MP][NP]=
        {0.0,1.0,1.0,3.0,-0.5,
        740.0,-1.0,0.0,-2.0,0.0,
        0.0,0.0,-2.0,0.0,7.0,
        0.5,0.0,-1.0,1.0,-2.0,
        9.0,-1.0,-1.0,-1.0,-1.0,
        0.0,0.0,0.0,0.0,0.0};
    float **a;
    static char *txt[NM1M2+1]=
        {" ","x1","x2","x3","x4","y1","y2","y3"};

    izrov=ivector(1,N);
    iposv=ivector(1,M);
    a=convert_matrix(&c[0][0],1,MP,1,NP);
    simplx(a,M,N,M1,M2,M3,&icase,izrov,iposv);
    if (icase == 1)
        printf("\nunbounded objective function\n");
    else if (icase == -1)
        printf("\nno solutions satisfy constraints given\n");
    else {
        printf("\n%11s"," ");
        for (i=1;i<=N;i++)
            if (izrov[i] <= NM1M2) printf("%10s",txt[izrov[i]]);
```

```
    printf("\n");
    for (i=1;i<=M+1;i++) {
        if (i > 1)
            printf("%s",txt[iposv[i-1]]);
        else
            printf("   ");
        printf("%10.2f",a[i][1]);
        for (j=2;j<=N+1;j++)
            if (izrov[j-1] <= NM1M2)
                printf("%10.2f",a[i][j]);
        printf("\n");
    }
}
free_convert_matrix(a,1,MP,1,NP);
free_ivector(iposv,1,M);
free_ivector(izrov,1,N);
}
```

anneal is a procedure for solving the travelling salesman problem—a problem that is included as a demonstration of the use of simulated annealing. Sample program xanneal.c has the function of setting up the initial route for the salesman and printing final results. ·For each of ncity=10 cities, it chooses random coordinates x[i],y[i] using routine ran3, and puts an entry for each city in the array iptr[i]. The array indicates the order in which the cities will be visited. On the originally specified path, the cities are in the order i=1,..,10 so the sample program initially takes iptr[i]=i. (It is assumed that the salesman will return to the first city after visiting the last.) A call is then made to anneal, which attempts to find the shortest alternative route, which is recorded in the array. After finding a path that resists further improvement, the driver lists the modified itinerary.

```
/* Driver for routine ANNEAL */

#include <stdio.h>
#include "nr.h"
#include "nrutil.h"

#define NCITY 10

main()
{
    int idum=(-111),i,ii,*iorder;
    float *x,*y;

    iorder=ivector(1,NCITY);
    x=vector(1,NCITY);
    y=vector(1,NCITY);
    for (i=1;i<=NCITY;i++) {
        x[i]=ran3(&idum);
        y[i]=ran3(&idum);
        iorder[i]=i;
    }
    anneal(x,y,iorder,NCITY);
    printf("*** System Frozen ***\n");
    printf("Final path:\n");
    printf("%8s %9s %12s\n","city","x","y");
    for (i=1;i<=NCITY;i++) {
```

```
        ii=iorder[i];
        printf("%4d %10.4f %10.4f\n",ii,x[ii],y[ii]);
    }
    free_vector(y,1,NCITY);
    free_vector(x,1,NCITY);
    free_ivector(iorder,1,NCITY);
}
```

Chapter 11: Eigensystems

In Chapter 11 of Numerical Recipes, we deal with the problem of finding eigenvectors and eigenvalues of matrices, first dealing with symmetric matrices, and then with more general cases. For real symmetric matrices of small-to-moderate size, the routine jacobi *is recommended as a simple and foolproof scheme of finding eigenvalues and eigenvectors. Routine* eigsrt *may be used to reorder the output of* jacobi *into ascending order of eigenvalue. A more efficient (but operationally more complicated) procedure is to reduce the symmetric matrix to tridiagonal form before doing the eigenvalue analysis.* tred2 *uses the Householder scheme to perform this reduction and is used in conjunction with* tqli. tqli *determines the eigenvalues and eigenvectors of a real, symmetric, tridiagonal matrix.*

For nonsymmetric matrices, we offer only routines for finding eigenvalues, and not eigenvectors. To ameliorate problems with roundoff error, balanc *makes the corresponding rows and columns of the matrix have comparable norms while leaving eigenvalues unchanged. Then the matrix is reduced to Hessenberg form by Gaussian elimination using* elmhes. *Finally* hqr *applies the QR algorithm to find the eigenvalues of the Hessenberg matrix.*

★   ★   ★   ★

jacobi is a reliable scheme for finding both the eigenvalues and eigenvectors of a symmetric matrix. It is not the most efficient scheme available, but it is simple and trustworthy, and it is recommended for problems of small-to-moderate order. Sample program xjacobi.c defines three matrices a, b, c. They are of order n=3, 5, and 10 respectively and are, each in turn, sent to jacobi. For each matrix, the eigenvalues and eigenvectors are reported. Then, an eigenvector test takes place in which the original matrix is applied to the purported eigenvector, and the ratio of the result to the vector itself is found. The ratio should, of course, be the eigenvalue.

```
/* Driver for routine JACOBI */

#include <stdio.h>
#include "nr.h"
#include "nrutil.h"

#define NP 10
#define NMAT 3

main()
{
    int i,j,k,kk,l,ll,nrot;
```

```
static float a[3][3]=
    {1.0,2.0,3.0,
    2.0,2.0,3.0,
    3.0,3.0,3.0};
static float b[5][5]=
    {-2.0,-1.0,0.0,1.0,2.0,
    -1.0,-1.0,0.0,1.0,2.0,
    0.0,0.0,0.0,1.0,2.0,
    1.0,1.0,1.0,1.0,2.0,
    2.0,2.0,2.0,2.0,2.0};
static float c[NP][NP]=
    {5.0,4.0,3.0,2.0,1.0,0.0,-1.0,-2.0,-3.0,-4.0,
    4.0,5.0,4.0,3.0,2.0,1.0,0.0,-1.0,-2.0,-3.0,
    3.0,4.0,5.0,4.0,3.0,2.0,1.0,0.0,-1.0,-2.0,
    2.0,3.0,4.0,5.0,4.0,3.0,2.0,1.0,0.0,-1.0,
    1.0,2.0,3.0,4.0,5.0,4.0,3.0,2.0,1.0,0.0,
    0.0,1.0,2.0,3.0,4.0,5.0,4.0,3.0,2.0,1.0,
    -1.0,0.0,1.0,2.0,3.0,4.0,5.0,4.0,3.0,2.0,
    -2.0,-1.0,0.0,1.0,2.0,3.0,4.0,5.0,4.0,3.0,
    -3.0,-2.0,-1.0,0.0,1.0,2.0,3.0,4.0,5.0,4.0,
    -4.0,-3.0,-2.0,-1.0,0.0,1.0,2.0,3.0,4.0,5.0};
float *d,*r,**v,**e;
static int num[4]={0,3,5,10};

d=vector(1,NP);
r=vector(1,NP);
v=matrix(1,NP,1,NP);
for (i=1;i<=NMAT;i++) {
    if (i == 1) e=convert_matrix(&a[0][0],1,num[i],1,num[i]);
    else if (i == 2) e=convert_matrix(&b[0][0],1,num[i],1,num[i]);
    else if (i == 3) e=convert_matrix(&c[0][0],1,num[i],1,num[i]);
    jacobi(e,num[i],d,v,&nrot);
    printf("matrix number %2d\n",i);
    printf("number of JACOBI rotations: %3d\n",nrot);
    printf("eigenvalues: \n");
    for (j=1;j<=num[i];j++) {
        printf("%12.6f",d[j]);
        if ((j % 5) == 0) printf("\n");
    }
    printf("\neigenvectors:\n");
    for (j=1;j<=num[i];j++) {
        printf("%9s %3d \n","number",j);
        for (k=1;k<=num[i];k++) {
            printf("%12.6f",v[k][j]);
            if ((k % 5) == 0) printf("\n");
        }
        printf("\n");
    }
    /* eigenvector test */
    printf("eigenvector test\n");
    for (j=1;j<=num[i];j++) {
        for (l=1;l<=num[i];l++) {
            r[l]=0.0;
            for (k=1;k<=num[i];k++) {
                if (k > l) {
                    kk=l;
                    ll=k;
```

```
                    } else {
                        kk=k;
                        ll=l;
                    }
                    r[l] += (e[ll][kk]*v[k][j]);
                }
            }
            printf("vector number %3d\n",j);
            printf("%11s %14s %10s\n",
                "vector","mtrx*vec.","ratio");
            for (l=1;l<=num[i];l++)
                printf("%12.6f %12.6f %12.6f\n",
                    v[l][j],r[l],r[l]/v[l][j]);
        }
        printf("press RETURN to continue...\n");
        getchar();
        free_convert_matrix(e,1,num[i],1,num[i]);
    }
    free_matrix(v,1,NP,1,NP);
    free_vector(r,1,NP);
    free_vector(d,1,NP);
}
```

eigsrt reorders the output of jacobi so that the eigenvectors are in the order of increasing eigenvalue. Sample program xeigsrt.c uses matrix c from the previous program to illustrate. This 10×10 matrix is passed to jacobi and the ten eigenvectors are found. They are printed, along with their eigenvalues, in the order that jacobi returns them. Then the matrices d and v from jacobi, which contain the eigenvalues and eigenvectors, are passed to eigsrt, and ought to return in ascending order of eigenvalue. The result is printed for inspection.

```
/* Driver for routine EIGSRT */

#include <stdio.h>
#include "nr.h"
#include "nrutil.h"

#define NP 10

main()
{
    int i,j,nrot;
    static float c[NP][NP]=
        {5.0,4.0,3.0,2.0,1.0,0.0,-1.0,-2.0,-3.0,-4.0,
         4.0,5.0,4.0,3.0,2.0,1.0,0.0,-1.0,-2.0,-3.0,
         3.0,4.0,5.0,4.0,3.0,2.0,1.0,0.0,-1.0,-2.0,
         2.0,3.0,4.0,5.0,4.0,3.0,2.0,1.0,0.0,-1.0,
         1.0,2.0,3.0,4.0,5.0,4.0,3.0,2.0,1.0,0.0,
         0.0,1.0,2.0,3.0,4.0,5.0,4.0,3.0,2.0,1.0,
         -1.0,0.0,1.0,2.0,3.0,4.0,5.0,4.0,3.0,2.0,
         -2.0,-1.0,0.0,1.0,2.0,3.0,4.0,5.0,4.0,3.0,
         -3.0,-2.0,-1.0,0.0,1.0,2.0,3.0,4.0,5.0,4.0,
         -4.0,-3.0,-2.0,-1.0,0.0,1.0,2.0,3.0,4.0,5.0};
    float *d,**v,**e;

    d=vector(1,NP);
    v=matrix(1,NP,1,NP);
```

```
    e=convert_matrix(&c[0][0],1,NP,1,NP);
    printf("****** Finding Eigenvectors ******\n");
    jacobi(e,NP,d,v,&nrot);
    printf("unsorted eigenvectors:\n");
    for (i=1;i<=NP;i++) {
        printf("eigenvalue %3d = %12.6f\n",i,d[i]);
        printf("eigenvector:\n");
        for (j=1;j<=NP;j++) {
            printf("%12.6f",v[j][i]);
            if ((j % 5) == 0) printf("\n");
        }
        printf("\n");
    }
    printf("\n****** Sorting Eigenvectors ******\n\n");
    eigsrt(d,v,NP);
    printf("sorted eigenvectors:\n");
    for (i=1;i<=NP;i++) {
        printf("eigenvalue %3d = %12.6f\n",i,d[i]);
        printf("eigenvector:\n");
        for (j=1;j<=NP;j++) {
            printf("%12.6f",v[j][i]);
            if ((j % 5) == 0) printf("\n");
        }
        printf("\n");
    }
    free_convert_matrix(e,1,NP,1,NP);
    free_matrix(v,1,NP,1,NP);
    free_vector(d,1,NP);
}
```

`tred2` reduces a real symmetric matrix to tridiagonal form. Sample program `tred2` again uses matrix c from the earlier programs, and copies it into matrix a. Matrix a is sent to `tred2`, while c is saved for a check of the transformation matrix that `tred2` returns in a. The program prints the diagonal and off-diagonal elements of the reduced matrix. It then forms the matrix f defined by $f = a^T c a$ to prove that f is tridiagonal and that the listed diagonal and off-diagonal elements are correct.

```
/* Driver for routine TRED2 */

#include <stdio.h>
#include "nr.h"
#include "nrutil.h"

#define NP 10

main()
{
    int i,j,k,l,m;
    float *d,*e,**a,**f;
    static float c[NP][NP]=
        { 5.0, 4.0, 3.0, 2.0, 1.0, 0.0,-1.0,-2.0,-3.0,-4.0,
          4.0, 5.0, 4.0, 3.0, 2.0, 1.0, 0.0,-1.0,-2.0,-3.0,
          3.0, 4.0, 5.0, 4.0, 3.0, 2.0, 1.0, 0.0,-1.0,-2.0,
          2.0, 3.0, 4.0, 5.0, 4.0, 3.0, 2.0, 1.0, 0.0,-1.0,
          1.0, 2.0, 3.0, 4.0, 5.0, 4.0, 3.0, 2.0, 1.0, 0.0,
          0.0, 1.0, 2.0, 3.0, 4.0, 5.0, 4.0, 3.0, 2.0, 1.0,
         -1.0, 0.0, 1.0, 2.0, 3.0, 4.0, 5.0, 4.0, 3.0, 2.0,
```

```
            -2.0,-1.0, 0.0, 1.0, 2.0, 3.0, 4.0, 5.0, 4.0, 3.0,
            -3.0,-2.0,-1.0, 0.0, 1.0, 2.0, 3.0, 4.0, 5.0, 4.0,
            -4.0,-3.0,-2.0,-1.0, 0.0, 1.0, 2.0, 3.0, 4.0, 5.0};

    d=vector(1,NP);
    e=vector(1,NP);
    a=matrix(1,NP,1,NP);
    f=matrix(1,NP,1,NP);
    for (i=1;i<=NP;i++)
        for (j=1;j<=NP;j++) a[i][j]=c[i-1][j-1];
    tred2(a,NP,d,e);
    printf("diagonal elements\n");
    for (i=1;i<=NP;i++) {
        printf("%12.6f",d[i]);
        if ((i % 5) == 0) printf("\n");
    }
    printf("off-diagonal elements\n");
    for (i=2;i<=NP;i++) {
        printf("%12.6f",e[i]);
        if ((i % 5) == 0) printf("\n");
    }
    /* Check transformation matrix */
    for (j=1;j<=NP;j++) {
        for (k=1;k<=NP;k++) {
            f[j][k]=0.0;
            for (l=1;l<=NP;l++) {
                for (m=1;m<=NP;m++)
                    f[j][k] += a[l][j]*c[l-1][m-1]*a[m][k];
            }
        }
    }
    /* How does it look? */
    printf("tridiagonal matrix\n");
    for (i=1;i<=NP;i++) {
        for (j=1;j<=NP;j++) printf("%7.2f",f[i][j]);
        printf("\n");
    }
    free_matrix(f,1,NP,1,NP);
    free_matrix(a,1,NP,1,NP);
    free_vector(e,1,NP);
    free_vector(d,1,NP);
}
```

tqli finds the eigenvectors and eigenvalues for a real, symmetric, tridiagonal matrix. Sample program xtqli.c operates with matrix c again, and uses tred2 to reduce it to tridiagonal form as before. More specifically, c is copied into matrix a, which is sent to tred2. From tred2 come two vectors d, e which are the diagonal and subdiagonal elements of the tridiagonal matrix. d and e are made arguments of tqli, as is a, the returned transformation matrix from tred2. On output from tqli, d is replaced with eigenvalues, and a with corresponding eigenvectors. These are checked as in the program for jacobi. That is, the original matrix c is applied to each eigenvector, and the result is divided (element by element) by the eigenvector. Look for a result equal to the eigenvalue. (Note: in some cases, the vector element is zero or nearly so. These cases are flagged with the words "div. by zero".)

```
/* Driver for routine TQLI */

#include <stdio.h>
#include <math.h>
#include "nr.h"
#include "nrutil.h"

#define NP 10
#define TINY 1.0e-6

main()
{
    int i,j,k;
    float *d,*e,*f,**a;
    static float c[NP][NP]=
        { 5.0, 4.0, 3.0, 2.0, 1.0, 0.0,-1.0,-2.0,-3.0,-4.0,
          4.0, 5.0, 4.0, 3.0, 2.0, 1.0, 0.0,-1.0,-2.0,-3.0,
          3.0, 4.0, 5.0, 4.0, 3.0, 2.0, 1.0, 0.0,-1.0,-2.0,
          2.0, 3.0, 4.0, 5.0, 4.0, 3.0, 2.0, 1.0, 0.0,-1.0,
          1.0, 2.0, 3.0, 4.0, 5.0, 4.0, 3.0, 2.0, 1.0, 0.0,
          0.0, 1.0, 2.0, 3.0, 4.0, 5.0, 4.0, 3.0, 2.0, 1.0,
         -1.0, 0.0, 1.0, 2.0, 3.0, 4.0, 5.0, 4.0, 3.0, 2.0,
         -2.0,-1.0, 0.0, 1.0, 2.0, 3.0, 4.0, 5.0, 4.0, 3.0,
         -3.0,-2.0,-1.0, 0.0, 1.0, 2.0, 3.0, 4.0, 5.0, 4.0,
         -4.0,-3.0,-2.0,-1.0, 0.0, 1.0, 2.0, 3.0, 4.0, 5.0};

    d=vector(1,NP);
    e=vector(1,NP);
    f=vector(1,NP);
    a=matrix(1,NP,1,NP);
    for (i=1;i<=NP;i++)
        for (j=1;j<=NP;j++) a[i][j]=c[i-1][j-1];
    tred2(a,NP,d,e);
    tqli(d,e,NP,a);
    printf("\nEigenvectors for a real symmetric matrix\n");
    for (i=1;i<=NP;i++) {
        for (j=1;j<=NP;j++) {
            f[j]=0.0;
            for (k=1;k<=NP;k++)
                f[j] += (c[j-1][k-1]*a[k][i]);
        }
        printf("%s %3d %s %10.6f\n","eigenvalue",i," =",d[i]);
        printf("%11s %14s %9s\n","vector","mtrx*vect.","ratio");
        for (j=1;j<=NP;j++) {
            if (fabs(a[j][i]) < TINY)
                printf("%12.6f %12.6f %12s\n",
                    a[j][i],f[j],"div. by 0");
            else
                printf("%12.6f %12.6f %12.6f\n",
                    a[j][i],f[j],f[j]/a[j][i]);
        }
        printf("Press ENTER to continue...\n");
        getchar();
    }
    free_matrix(a,1,NP,1,NP);
    free_vector(f,1,NP);
    free_vector(e,1,NP);
```

```
        free_vector(d,1,NP);
}
```

balanc reduces error in eigenvalue problems involving non-symmetric matrices. It does this by adjusting corresponding rows and columns to have comparable norms, without changing eigenvalues. Sample program xbalanc.c prepares the following array a for balanc

$$\begin{pmatrix} 1 & 100 & 1 & 100 & 1 \\ 1 & 1 & 1 & 1 & 1 \\ 1 & 100 & 1 & 100 & 1 \\ 1 & 1 & 1 & 1 & 1 \\ 1 & 100 & 1 & 100 & 1 \end{pmatrix}$$

The norms of the five rows and five columns are printed out. It is clear from the array that three of the rows and two of the columns have much larger norms than the others. After balancing with balanc, the norms are recalculated, and this time the row, column pairs should be much more nearly equal.

```
/* Driver for routine BALANC */

#include <stdio.h>
#include <math.h>
#include "nr.h"
#include "nrutil.h"

#define NP 5

main()
{
    int i,j;
    float *c,*r,**a;

    c=vector(1,NP);
    r=vector(1,NP);
    a=matrix(1,NP,1,NP);
    for (i=1;i<=NP;i++)
        for (j=1;j<=NP;j++) {
            if ((i%2 == 1) && (j%2 == 0))
                a[i][j] = 100.0;
            else
                a[i][j] = 1.0;
        }
    /* Write norms */
    for (i=1;i<=NP;i++) {
        r[i]=0.0;
        c[i]=0.0;
        for (j=1;j<=NP;j++) {
            r[i] += fabs(a[i][j]);
            c[i] += fabs(a[j][i]);
        }
    }
    printf("rows:\n");
    for (i=1;i<=NP;i++) printf("%12.2f",r[i]);
    printf("\ncolumns:\n");
    for (i=1;i<=NP;i++) printf("%12.2f",c[i]);
    printf("\n\n***** Balancing matrix *****\n\n");
```

```
    balanc(a,NP);
    /* Write norms */
    for (i=1;i<=NP;i++) {
        r[i]=0.0;
        c[i]=0.0;
        for (j=1;j<=NP;j++) {
            r[i] += fabs(a[i][j]);
            c[i] += fabs(a[j][i]);
        }
    }
    printf("rows:\n");
    for (i=1;i<=NP;i++) printf("%12.2f",r[i]);
    printf("\ncolumns:\n");
    for (i=1;i<=NP;i++) printf("%12.2f",c[i]);
    printf("\n");
    free_matrix(a,1,NP,1,NP);
    free_vector(r,1,NP);
    free_vector(c,1,NP);
}
```

elmhes reduces a general matrix to Hessenberg form using Gaussian elimination. It is particularly valuable for real, non-symmetric matrices. Sample program xelmhes.c employs balanc and elmhes to get a non-symmetric and grossly unbalanced matrix into Hessenberg form. The matrix a is

$$\begin{pmatrix} 1 & 2 & 300 & 4 & 5 \\ 2 & 3 & 400 & 5 & 6 \\ 3 & 4 & 5 & 6 & 7 \\ 4 & 5 & 600 & 7 & 8 \\ 5 & 6 & 700 & 8 & 9 \end{pmatrix}$$

After printing the original matrix, the program feeds it to balanc and prints the balanced version. This is submitted to elmhes and the result is printed. Notice that the elements of a with $i>j+1$ are all set to zero by the program, because elmhes returns random values in this part of the matrix. Therefore, you should not attach any importance to the fact that the printed output of the program has Hessenberg form. More important are the contents of the non-zero entries. We include here the expected results for comparison.

Balanced Matrix:

1.00	2.00	37.50	4.00	5.00
2.00	3.00	50.00	5.00	6.00
24.00	32.00	5.00	48.00	56.00
4.00	5.00	75.00	7.00	8.00
5.00	6.00	87.50	8.00	9.00

Reduced to Hessenberg Form:

.1000E+01	.3938E+02	.9618E+01	.3333E+01	.4000E+01
.2400E+02	.2733E+02	.1161E+03	.4800E+02	.4800E+02
.0000E+00	.8551E+02	-.4780E+01	-.1333E+01	-.2000E+01
.0000E+00	.0000E+00	.5188E+01	.1447E+01	.2171E+01
.0000E+00	.0000E+00	.0000E+00	-.9155E-07	.7874E-07

```
/* Driver for routine ELMHES */

#include <stdio.h>
#include "nr.h"
```

```
#include "nrutil.h"

#define NP 5

main()
{
    int i,j;
    static float b[NP][NP]=
        {1.0,2.0,300.0,4.0,5.0,
         2.0,3.0,400.0,5.0,6.0,
         3.0,4.0,5.0,6.0,7.0,
         4.0,5.0,600.0,7.0,8.0,
         5.0,6.0,700.0,8.0,9.0};
    float **a;

    a=convert_matrix(&b[0][0],1,NP,1,NP);
    printf("***** original matrix *****\n");
    for (i=1;i<=NP;i++) {
        for (j=1;j<=NP;j++) printf("%12.2f",a[i][j]);
        printf("\n");
    }
    printf("***** balance matrix *****\n");
    balanc(a,NP);
    for (i=1;i<=NP;i++) {
        for (j=1;j<=NP;j++) printf("%12.2f",a[i][j]);
        printf("\n");
    }
    printf("***** reduce to hessenberg form *****\n");
    elmhes(a,NP);
    for (j=1;j<=NP-2;j++)
        for (i=j+2;i<=NP;i++)
            a[i][j]=0.0;
    for (i=1;i<=NP;i++) {
        for (j=1;j<=NP;j++) printf("%12.2e",a[i][j]);
        printf("\n");
    }
    free_convert_matrix(a,1,NP,1,NP);
}
```

hqr, finally, is a routine for finding the eigenvalues of a Hessenberg matrix using the QR algorithm. The 5×5 matrix specified in the array a is treated just as you would expect to treat any general real non-symmetric matrix. It is fed to balanc for balancing, to elmhes for reduction to Hessenberg form, and to hqr for eigenvalue determination. The eigenvalues may be complex-valued, and both real and imaginary parts are given. The original matrix has enough strategically placed zeros in it that you should have no trouble finding the eigenvalues by hand. Alternatively, you may check them against the list below:

Matrix:

1.00	2.00	.00	.00	.00
-2.00	3.00	.00	.00	.00
3.00	4.00	50.00	.00	.00
-4.00	5.00	-60.00	7.00	.00
-5.00	6.00	-70.00	8.00	-9.00

Eigenvalues:

```
#         Real          Imag.
1     .500000E+02    .000000E+00
2     .200000E+01   -.173205E+01
3     .200000E+01    .173205E+01
4     .700000E+01    .000000E+00
5    -.900000E+01    .000000E+00
```

```c
/* driver for routine HQR */

#include <stdio.h>
#include "nr.h"
#include "nrutil.h"

#define NP 5

main()
{
    int i,j;
    static float c[NP][NP]=
        {1.0,2.0,0.0,0.0,0.0,
        -2.0,3.0,0.0,0.0,0.0,
        3.0,4.0,50.0,0.0,0.0,
        -4.0,5.0,-60.0,7.0,0.0,
        -5.0,6.0,-70.0,8.0,-9.0};
    float *wr,*wi,**a;

    wr=vector(1,NP);
    wi=vector(1,NP);
    a=convert_matrix(&c[0][0],1,NP,1,NP);
    printf("matrix:\n");
    for (i=1;i<=NP;i++) {
        for (j=1;j<=NP;j++) printf("%12.2f",a[i][j]);
        printf("\n");
    }
    balanc(a,NP);
    elmhes(a,NP);
    hqr(a,NP,wr,wi);
    printf("eigenvalues:\n");
    printf("%11s %16s \n","real","imag.");
    for (i=1;i<=NP;i++) printf("%15f %14f\n",wr[i],wi[i]);
    free_convert_matrix(a,1,NP,1,NP);
    free_vector(wi,1,NP);
    free_vector(wr,1,NP);
}
```

Chapter 12: Fourier Methods

Chapter 12 of Numerical Recipes covers Fourier transform spectral methods, particularly the transform of discretely sampled data. Central to the chapter is the fast Fourier transform (FFT). Routine four1 *performs the FFT on a complex data array.* twofft *does the same transform on two real-valued data arrays (at the same time) and returns two complex-valued transforms. Finally,* realft *finds the Fourier transform of a single real-valued array. Two related transforms are the sine transform and the cosine transform, given by* sinft *and* cosft.

Two common uses of the Fourier transform are the convolution of data with a response function, and the computation of the correlation of two data sets. These operations are carried out by convlv *and* correl *respectively. Other applications of Fourier methods include data filtering, power spectrum estimation (*spctrm*, or* evlmem *with* memcof*), and linear prediction (*predic *with* fixrts*). All of these applications assume data in one dimension. For FFTs in two or more dimensions the routine* fourn *is supplied.*

★   ★   ★   ★

Routine four1 performs the fast Fourier transform on a complex-valued array of data points. Example program xfour1.c has five tests for this transform. First, it checks the following four symmetries (where $h(t)$ is the data and $H(n)$ is the transform):

1. If $h(t)$ is real-valued and even, then $H(n) = H(N - n)$ and H is real.

2. If $h(t)$ is imaginary-valued and even, then $H(n) = H(N - n)$ and H is imaginary.

3. If $h(t)$ is real-valued and odd, then $H(n) = -H(N - n)$ and H is imaginary.

4. If $h(t)$ is imaginary-valued and odd, then $H(n) = -H(N - n)$ and H is real.

The fifth test is that if a data array is Fourier transformed twice in succession, the resulting array should be identical to the original.

```
/* Driver for routine FOUR1 */

#include <stdio.h>
#include <math.h>
#include "nr.h"
#include "nrutil.h"

void prntft(data,nn)
float data[];
int nn;
```

150

```
{
    int n;

    printf("%4s %13s %13s %12s %13s\n",
        "n","real(n)","imag.(n)","real(N-n)","imag.(N-n)");
    printf("    0 %14.6f %12.6f %12.6f %12.6f\n",
        data[1],data[2],data[1],data[2]);
    for (n=3;n<=nn+1;n += 2) {
        printf("%4d %14.6f %12.6f %12.6f %12.6f\n",
            ((n-1)/2),data[n],data[n+1],
            data[2*nn+2-n],data[2*nn+3-n]);
    }
    printf(" press return to continue ...\n");
    getchar();
    return;
}


#define NN 32
#define NN2 (2*NN)
#define SQR(a) ((a)*(a))

main()
{
    int i,isign;
    float *data,*dcmp;

    data=vector(1,NN2);
    dcmp=vector(1,NN2);
    printf("h(t)=real-valued even-function\n");
    printf("h(n)=h(N-n) and real?\n");
    for (i=1;i<NN2;i += 2) {
        data[i]=1.0/(SQR((float) (i-NN-1)/NN)+1.0);
        data[i+1]=0.0;
    }
    isign=1;
    four1(data,NN,isign);
    prntft(data,NN);
    printf("h(t)=imaginary-valued even-function\n");
    printf("h(n)=h(N-n) and imaginary?\n");
    for (i=1;i<NN2;i += 2) {
        data[i+1]=1.0/(SQR((float) (i-NN-1)/NN)+1.0);
        data[i]=0.0;
    }
    isign=1;
    four1(data,NN,isign);
    prntft(data,NN);
    printf("h(t)=real-valued odd-function\n");
    printf("h(n) = -h(N-n) and imaginary?\n");
    for (i=1;i<NN2;i += 2) {
        data[i]=((float) (i-NN-1)/NN)/(SQR((float) (i-NN-1)/NN)+1.0);
        data[i+1]=0.0;
    }
    data[1]=0.0;
    isign=1;
    four1(data,NN,isign);
    prntft(data,NN);
    printf("h(t)=imaginary-valued odd-function\n");
```

```
        printf("h(n) = -h(N-n) and real?\n");
        for (i=1;i<NN2;i += 2) {
            data[i+1]=((float) (i-NN-1)/NN)/(SQR((float) (i-NN-1)/NN)+1.0);
            data[i]=0.0;
        }
        data[2]=0.0;
        isign=1;
        four1(data,NN,isign);
        prntft(data,NN);
        /* transform, inverse-transform test */
        for (i=1;i<NN2;i += 2) {
            data[i]=1.0/(SQR(0.5*(i-NN-1.0)/NN)+1.0);
            dcmp[i]=data[i];
            data[i+1]=(0.25*(i-NN-1.0)/NN)*exp(-SQR(0.5*(i-NN-1)/NN));
            dcmp[i+1]=data[i+1];
        }
        isign=1;
        four1(data,NN,isign);
        isign = -1;
        four1(data,NN,isign);
        printf("%33s %23s \n","double fourier transform:","original data:");
        printf("\n %3s %15s %12s %12s %12s \n",
            "k","real h(k)","imag h(k)","real h(k)","imag h(k)");
        for (i=1;i<NN;i += 2)
            printf("%4d %14.6f %12.6f %12.6f %12.6f\n",
                (i+1)/2,dcmp[i],dcmp[i+1],data[i]/NN,data[i+1]/NN);
        free_vector(dcmp,1,NN2);
        free_vector(data,1,NN2);
}
```

twofft is a routine that performs an efficient FFT of two real arrays at once by packing them into a complex array and transforming with four1. Sample program xtwofft.c generates two periodic data sets, out of phase with one another, and performs a transform and an inverse transform on each. It will be difficult to judge whether the transform itself gives the right answer, but if the inverse transform gets you back to the easily recognized original, you may be fairly confident that the routine works.

```
/* Driver for routine TWOFFT */

#include <stdio.h>
#include <math.h>
#include "nr.h"
#include "nrutil.h"

#define N 32
#define N2 (2*N)
#define PER 8
#define PI 3.1415926

void prntft(data,nn)
float data[];
int nn;
{
    int n;

    printf("%4s %13s %13s %12s %13s\n",
        "n","real(n)","imag.(n)","real(N-n)","imag.(N-n)");
```

```
        printf("    0 %14.6f %12.6f %12.6f %12.6f\n",
            data[1],data[2],data[1],data[2]);
        for (n=3;n<=nn+1;n += 2) {
            printf("%4d %14.6f %12.6f %12.6f %12.6f\n",
                ((n-1)/2),data[n],data[n+1],
                data[2*nn+2-n],data[2*nn+3-n]);
        }
        printf(" press return to continue ...\n");
        getchar();
        return;
}

main()
{
    int i,isign;
    float *data1,*data2,*fft1,*fft2;

    data1=vector(1,N);
    data2=vector(1,N);
    fft1=vector(1,N2);
    fft2=vector(1,N2);
    for (i=1;i<=N;i++) {
        data1[i]=floor(0.5+cos(i*2.0*PI/PER));
        data2[i]=floor(0.5+sin(i*2.0*PI/PER));
    }
    twofft(data1,data2,fft1,fft2,N);
    printf("Fourier transform of first function:\n");
    prntft(fft1,N);
    printf("Fourier transform of second function:\n");
    prntft(fft2,N);
    /* Invert transform */
    isign = -1;
    four1(fft1,N,isign);
    printf("inverted transform =  first function:\n");
    prntft(fft1,N);
    four1(fft2,N,isign);
    printf("inverted transform =  second function:\n");
    prntft(fft2,N);
    free_vector(fft2,1,N2);
    free_vector(fft1,1,N2);
    free_vector(data2,1,N);
    free_vector(data1,1,N);
}
```

realft performs the Fourier transform of a single real-valued data array. Sample routine xrealft.c takes this function to be sinusoidal, and allows you to choose the period. After transforming, it simply plots the magnitude of each element of the transform. If the period you choose is a power of two, the transform will be nonzero in a single bin; otherwise there will be leakage to adjacent channels. xrealft.c follows every transform by an inverse transform to make sure the original function is recovered.

```
/* Driver for routine REALFT */

#include <stdio.h>
#include <math.h>
#include "nr.h"
#include "nrutil.h"
```

```
#define EPS 1.0e-3
#define NP 32
#define WIDTH 50.0
#define PI 3.1415926
#define SQR(a) ((a)*(a))

main()
{
    int i,j,n=NP/2,nlim;
    float big,per,scal,small,*data,*size;

    data=vector(1,NP);
    size=vector(1,NP/2+1);
    for (;;) {
        printf("Period of sinusoid in channels (2-%2d)\n",NP);
        scanf("%f",&per);
        if (per <= 0.0) break;
        for (i=1;i<=NP;i++)
            data[i]=cos(2.0*PI*(i-1)/per);
        realft(data,n,1);
        big = -1.0e10;
        for (i=2;i<=n;i++) {
            size[i]=sqrt(SQR(data[2*i-1])+SQR(data[2*i]));
            if (size[i] > big) big=size[i];
        }
        size[1]=fabs(data[1]);
        if (size[1] > big) big=size[1];
        size[n+1]=fabs(data[2]);
        if (size[n+1] > big) big=size[n+1];
        scal=WIDTH/big;
        for (i=1;i<=n+1;i++) {
            nlim=(int) (0.5+scal*size[i]+EPS);
            printf("%4d ",i);
            for (j=1;j<=nlim+1;j++) printf("*");
            printf("\n");
        }
        printf("press RETURN to continue ...\n");
        getchar();
        realft(data,n,-1);
        big = -1.0e10;
        small=1.0e10;
        for (i=1;i<=NP;i++) {
            if (data[i] < small) small=data[i];
            if (data[i] > big) big=data[i];
        }
        scal=WIDTH/(big-small);
        for (i=1;i<=NP;i++) {
            nlim=(int) (0.5+scal*(data[i]-small)+EPS);
            printf("%4d ",i);
            for (j=1;j<=nlim+1;j++) printf("*");
            printf("\n");
        }
    }
    free_vector(size,1,NP/2+1);
    free_vector(data,1,NP);
}
```

sinft performs a sine-transform of a real-valued array. The necessity for such a transform arises in solution methods for partial differential equations with certain kinds of boundary conditions (see Chapter 17). The sample program xsinft.c works exactly as the previous program. Notice that in this program no distinction needs to be made between the transform and its inverse. They are identical.

```c
/* Driver for routine SINFT */

#include <stdio.h>
#include <math.h>
#include "nr.h"
#include "nrutil.h"

#define EPS 1.0e-3
#define NP 16
#define WIDTH 30.0
#define PI 3.1415926

main()
{
    float big,per,scal,small,*data;
    int i,j,nlim;

    data=vector(1,NP);
    for (;;) {
        printf("\nPeriod of sinusoid in channels (3-%2d)\n",NP);
        scanf("%f",&per);
        if (per <= 0.0) break;
        for (i=1;i<=NP;i++)
            data[i]=sin(2.0*PI*(i-1)/per);
        sinft(data,NP);
        big = -1.0e10;
        small=1.0e10;
        for (i=1;i<=NP;i++) {
            if (data[i] < small) small=data[i];
            if (data[i] > big) big=data[i];
        }
        scal=WIDTH/(big-small);
        for (i=1;i<=NP;i++) {
            nlim=(int) (scal*(data[i]-small)+EPS+0.5);
            printf("%4d  ",i);
            for (j=1;j<=nlim+1;j++) printf("*");
            printf("\n");
        }
        printf("press RETURN to continue ...\n");
        getchar();
        sinft(data,NP);
        big = -1.0e10;
        small=1.0e10;
        for (i=1;i<=NP;i++) {
            if (data[i] < small) small=data[i];
            if (data[i] > big) big=data[i];
        }
        scal=WIDTH/(big-small);
        for (i=1;i<=NP;i++) {
            nlim=(int) (scal*(data[i]-small)+EPS+0.5);
            printf("%4d  ",i);
```

```
            for (j=1;j<=nlim+1;j++) printf("*");
            printf("\n");
        }
    }
    free_vector(data,1,NP);
}
```

cosft is a companion procedure to sinft that does the cosine transform. It also plays a role in partial differential equation solutions. Although program xcosft.c is again the same as xrealft.c, you will notice some difference in solutions. The cosine transform of a cosine with a period that is a power of two does not give a transform that is nonzero in a single bin. It has some small values at other frequencies. This is due to our desire to cast the transform into something that calls realft, and therefore works on 2^N points rather than the more natural $2^N + 1$. The sample program will prove to you, however, that the transform expressed here is invertible. Notice that, unlike the sine transform, the cosine transform is not self-inverting.

```
/* Driver for routine COSFT */

#include <stdio.h>
#include <math.h>
#include "nr.h"
#include "nrutil.h"

#define EPS 1.0e-3
#define NP 16
#define WIDTH 30.0
#define PI 3.1415926

main()
{
    int i,j,nlim;
    float big,per,scal,small,*data;

    data=vector(1,NP);
    for (;;) {
        printf("Period of cosine in channels (2-%2d)\n",NP);
        scanf("%f",&per);
        if (per <= 0.0) break;
        for (i=1;i<=NP;i++)
            data[i]=cos(2.0*PI*(i-1)/per);
        cosft(data,NP,1);
        big = -1.0e10;
        small=1.0e10;
        for (i=1;i<=NP;i++) {
            if (data[i] < small) small=data[i];
            if (data[i] > big) big=data[i];
        }
        scal=WIDTH/(big-small);
        for (i=1;i<=NP;i++) {
            nlim=(int) (0.5+scal*(data[i]-small)+EPS);
            printf("%4d ",i);
            for (j=1;j<=nlim+1;j++) printf("*");
            printf("\n");
        }
        printf("press RETURN to continue ...\n");
```

```
        getchar();
        cosft(data,NP,-1);
        big = -1.0e10;
        small=1.0e10;
        for (i=1;i<=NP;i++) {
            if (data[i] < small) small=data[i];
            if (data[i] > big) big=data[i];
        }
        scal=WIDTH/(big-small);
        for (i=1;i<=NP;i++) {
            nlim=(int) (0.5+scal*(data[i]-small)+EPS);
            printf("%4d ",i);
            for (j=1;j<=nlim+1;j++) printf("*");
            printf("\n");
        }
    }
    free_vector(data,1,NP);
}
```

Procedure `convlv` performs the convolution of a data set with a response function using an FFT. Sample program `xconvlv.c` uses two functions that take on only the values 0.0 and 1.0. The data array `data[i]` has sixteen values, and is zero everywhere except between `i=6` and `i=10` where it is 1.0. The response function `respns[i]` has nine values and is zero except between `i=3` and `i=6` where it is 1.0. The expected value of the convolution is determined simply by flipping the response function end-to-end, moving it to the left by the desired shift, and counting how many non-zero channels of `respns` fall on non-zero channels of `data`. In this way, you should be able to verify the result from the program. The sample program, incidentally, does the calculation by this direct method for the purpose of comparison.

```
/* Driver for routine CONVLV */

#include <stdio.h>
#include "nr.h"
#include "nrutil.h"

#define N 16          /* data array size */
#define M 9           /* response function dimension - must be odd */
#define N2 (2*N)

main()
{
    int i,isign,j;
    float cmp,*data,*respns,*resp,*ans;

    data=vector(1,N);
    respns=vector(1,N);
    resp=vector(1,N);
    ans=vector(1,N2);
    for (i=1;i<=N;i++)
        if ((i >= N/2-N/8) && (i <= N/2+N/8))
            data[i]=1.0;
        else
            data[i]=0.0;
    for (i=1;i<=M;i++) {
        if ((i > 2) && (i < 7))
```

```
                    respns[i]=1.0;
            else
                    respns[i]=0.0;
            resp[i]=respns[i];
    }
    isign=1;
    convlv(data,N,resp,M,isign,ans);
    /* compare with a direct convolution */
    printf("%3s %14s %13s\n","i","CONVLV","Expected");
    for (i=1;i<=N;i++) {
        cmp=0.0;
        for (j=1;j<=M/2;j++) {
            cmp += data[((i-j-1+N) % N)+1]*respns[j+1];
            cmp += data[((i+j-1) % N)+1]*respns[M-j+1];
        }
        cmp += data[i]*respns[1];
        printf("%3d %15.6f %12.6f\n",i,ans[i],cmp);
    }
    free_vector(ans,1,N2);
    free_vector(resp,1,N);
    free_vector(respns,1,N);
    free_vector(data,1,N);
}
```

correl calculates the correlation function of two data sets. Sample program xcor-rel.c defines data1[i] as an array of 64 values which are all zero except from i=25 to i=39, where they are one. data2[i] is defined in the same way. Therefore, the correlation being performed is an autocorrelation. The sample routine compares the result of the calculation as performed by correl with that found by a direct calculation. In this case the calculation may be done manually simply by successively shifting data2 with respect to data1 and counting the number of nonzero channels of the two that overlap.

```
/* Driver for routine CORREL */

#include <stdio.h>
#include "nr.h"
#include "nrutil.h"

#define N 64
#define N2 (2*N)

main()
{
    int i,j;
    float cmp,*data1,*data2,*ans;

    data1=vector(1,N);
    data2=vector(1,N);
    ans=vector(1,N2);
    for (i=1;i<=N;i++) {
        if ((i > N/2-N/8) && (i < N/2+N/8))
            data1[i]=1.0;
        else
            data1[i]=0.0;
        data2[i]=data1[i];
    }
```

```
        correl(data1,data2,N,ans);
        /* Calculate directly */
        printf("%3s %14s %18s\n","n","CORREL","direct calc.");
        for (i=0;i<=16;i++) {
            cmp=0.0;
            for (j=1;j<=N;j++)
                cmp += data1[((i+j-1) % N)+1]*data2[j];
            printf("%3d %15.6f %15.6f\n",i,ans[i+1],cmp);
        }
        free_vector(ans,1,N2);
        free_vector(data2,1,N);
        free_vector(data1,1,N);
}
```

spctrm does a spectral estimate of a data set by reading it in as segments, windowing, Fourier transforming, and accumulating the power spectrum. Data segments may or may not be overlapped at the decision of the user. In sample program xspctrm.c the spectral data is read in from a file called spctrl.dat containing 1200 numbers and included on the *Numerical Recipes Examples Diskette*. It is analyzed first with overlap and then without. The results are tabulated side by side for comparison.

```
/* Driver for routine SPCTRM */

#include <stdio.h>
#include "nr.h"
#include "nrutil.h"

#define M 16
#define TRUE 1
#define FALSE 0

main()
{
    int j,k,ovrlap;
    float *p,*q;
    FILE *fp;

    p=vector(1,M);
    q=vector(1,M);
    if ((fp = fopen("spctrl.dat","r")) == NULL)
        nrerror("Data file SPCTRL.DAT not found\n");
    k=8;
    ovrlap=TRUE;
    spctrm(fp,p,M,k,ovrlap);
    rewind(fp);
    k=16;
    ovrlap=FALSE;
    spctrm(fp,q,M,k,ovrlap);
    fclose(fp);
    printf("\nSpectrum of data in file SPCTRL.DAT\n");
    printf("%13s %s %5s %s\n"," ","overlapped "," ","non-overlapped");
    for (j=1;j<=M;j++)
        printf("%3d %5s %13f %5s %13f\n",j," ",p[j]," ",q[j]);
    free_vector(q,1,M);
    free_vector(p,1,M);
}
```

memcof and evlmem are used to perform spectral analysis by the maximum entropy method. memcof finds the coefficients for a model spectrum, the magnitude squared of the inverse of a polynomial series. Sample program xmemcof.c determines the coefficients for 1000 numbers from the file spctrl.dat and simply prints the results for comparison to the following table:

```
Coefficients for spectral estimation of spctrl.dat
 a[ 1] =    1.261539
 a[ 2] =   -0.007695
 a[ 3] =   -0.646778
 a[ 4] =   -0.280603
 a[ 5] =    0.163693
 a[ 6] =    0.347674
 a[ 7] =    0.111247
 a[ 8] =   -0.337141
 a[ 9] =   -0.358043
 a[10] =    0.378774
    b0 =    0.003511
```

```
/* Driver for routine MEMCOF */

#include <stdio.h>
#include "nr.h"
#include "nrutil.h"

#define N 1000
#define M 10

main()
{
    int i;
    float pm,*cof,*data;
    FILE *fp;

    cof=vector(1,M);
    data=vector(1,N);
    if ((fp = fopen("spctrl.dat","r")) == NULL)
        nrerror("Data file SPCTRL.DAT not found\n");
    for (i=1;i<=N;i++) fscanf(fp,"%f",&data[i]);
    fclose(fp);
    memcof(data,N,M,&pm,cof);
    printf("Coefficients for spectral estimation of SPCTRL.DAT\n\n");
    for (i=1;i<=M;i++) printf("a[%2d] = %12.6f \n",i,cof[i]);
    printf("\nb0 =%12.6f\n",pm);
    free_vector(data,1,N);
    free_vector(cof,1,M);
}
```

evlmem uses coefficients from memcof to generate a spectral estimate. The example xevlmem.c uses the same data from spctrl.dat and prints the spectral estimate. You may compare the result to:

```
Power spectrum estimate of data in spctrl.dat
    f*delta        power
    0.000000     0.026023
    0.031250     0.029266
    0.062500     0.193087
```

```
0.093750      0.139241
0.125000     29.915518
0.156250      0.003878
0.187500      0.000633
0.218750      0.000334
0.250000      0.000437
0.281250      0.001331
0.312500      0.000780
0.343750      0.000451
0.375000      0.000784
0.406250      0.001381
0.437500      0.000649
0.468750      0.000775
0.500000      0.001716
```

```
/* Driver for routine EVLMEM */

#include <stdio.h>
#include "nr.h"
#include "nrutil.h"

#define N 1000
#define M 10
#define NFDT 16

main()
{
    int i;
    float fdt,pm,*cof,*data;
    FILE *fp;

    cof=vector(1,M);
    data=vector(1,N);
    if ((fp = fopen("spctrl.dat","r")) == NULL)
        nrerror("Data file SPCTRL.DAT not found\n");
    for (i=1;i<=N;i++) fscanf(fp,"%f",&data[i]);
    fclose(fp);
    memcof(data,N,M,&pm,cof);
    printf("Power spectum estimate of data in SPCTRL.DAT\n");
    printf("     f*delta         power\n");
    for (fdt=0.0;fdt<=0.5;fdt += 0.5/NFDT)
        printf("%12.6f %12.6f\n",fdt,evlmem(fdt,cof,M,pm));
    free_vector(data,1,N);
    free_vector(cof,1,M);
}
```

Notice that once memcof has determined coefficients, we may evaluate the estimate at any intervals we wish. Notice also that we have built a spectral peak into the noisy data in spctrl.dat.

Linear prediction is carried out by routines predic, memcof, and fixrts. memcof produces the linear prediction coefficients from the data set. fixrts massages the coefficients so that all roots of the characteristic polynomial fall inside the unit circle of the complex domain, thus insuring stability of the prediction algorithm. Finally, predic predicts future data points based on the modified coefficients. Sample program xfixrts.c demonstrates the operation of fixrts. The coefficients provided in the array d[i] are those appropriate to the polynomial $(z-1)^6 = 1$. This equation has six roots on a circle of radius one, centered at $(1.0, 0.0)$ in the complex plane. Some of these lie within the unit circle and some outside. The ones outside are moved by fixrts according to $z_i \rightarrow 1/z_i^*$. You can easily figure these out by hand and check the results. Also, the sample routine calculates $(z-1)^6$ for each of the adjusted roots, and thereby shows which have been changed and which have not.

```c
/* Driver for routine FIXRTS */

#include <stdio.h>
#include "nr.h"
#include "complex.h"

#define NPOLES 6
#define NP1 (NPOLES+1)
#define ONE Complex(1.0,0.0)
#define TRUE 1

main()
{
    int i,polish;
    static float d[NP1]=
        {0.0,6.0,-15.0,20.0,-15.0,6.0,0.0};
    fcomplex zcoef[NP1],zeros[NP1],z1,z2;

    polish=TRUE;
    /* finding roots of (z-1.0)^6=1.0 */
    /* first write roots */
    zcoef[NPOLES]=ONE;
    for (i=NPOLES-1;i>=0;i--)
        zcoef[i] = Complex(-d[NPOLES-i],0.0);
    zroots(zcoef,NPOLES,zeros,polish);
    printf("Roots of (z-1.0)^6 = 1.0\n");
    printf("%24s %27s \n","Root","(z-1.0)^6");
    for (i=1;i<=NPOLES;i++) {
        z1=Csub(zeros[i],ONE);
        z2=Cmul(z1,z1);
        z1=Cmul(z1,z2);
        z1=Cmul(z1,z1);
        printf("%6d %12.6f %12.6f %12.6f %12.6f\n",
            i,zeros[i].r,zeros[i].i,z1.r,z1.i);
    }
    /* now fix them to lie within unit circle */
    fixrts(d,NPOLES);
    /* check results */
    zcoef[NPOLES]=ONE;
    for (i=NPOLES-1;i>=0;i--)
        zcoef[i] = Complex(-d[NPOLES-i],0.0);
    zroots(zcoef,NPOLES,zeros,polish);
    printf("\nRoots reflected in unit circle\n");
    printf("%24s %27s \n","Root","(z-1.0)^6");
    for (i=1;i<=NPOLES;i++) {
        z1=Csub(zeros[i],ONE);
        z2=Cmul(z1,z1);
        z1=Cmul(z1,z2);
        z1=Cmul(z1,z1);
        printf("%6d %12.6f %12.6f %12.6f %12.6f\n",
```

```
                i,zeros[i].r,zeros[i].i,z1.r,z1.i);
        }
}
```

predic carries out the job of performing the prediction. The function chosen for investigation in sample program xpredic.c is

$$F(n) = \exp(-n/\mathrm{npts})\sin(2\pi n/50) + \exp(-2n/\mathrm{npts})\sin(2.2\pi n/50)$$

the sum of two sine waves of similar period and exponentially decaying amplitudes. On the basis of 300 data points, and working with coefficients representing ten poles, the routine predicts 20 future points. The quality of this prediction may be judged by comparing these 20 points with the evaluations of $F(n)$ that are provided.

```
/* Driver for routine PREDIC */

#include <stdio.h>
#include <math.h>
#include "nr.h"
#include "nrutil.h"

#define NPTS 300
#define NPOLES 10
#define NFUT 20
#define PI 3.1415926

float f(n,npts)
int n,npts;
{
    return exp(-(1.0*n)/npts)*sin(2.0*PI*n/50.0)
        +exp(-(2.0*n)/npts)*sin(2.2*PI*n/50.0);
}

main()
{
    int i;
    float dum,*d,*future,*data;

    d=vector(1,NPOLES);
    future=vector(1,NFUT);
    data=vector(1,NPTS);
    for (i=1;i<=NPTS;i++)
        data[i]=f(i,NPTS);
    memcof(data,NPTS,NPOLES,&dum,d);
    fixrts(d,NPOLES);
    predic(data,NPTS,d,NPOLES,future,NFUT);
    printf("%6s %11s %12s\n","I","Actual","PREDIC");
    for (i=1;i<=NFUT;i++)
        printf("%6d %12.6f %12.6f\n",i,f(i+NPTS,NPTS),future[i]);
    free_vector(data,1,NPTS);
    free_vector(future,1,NFUT);
    free_vector(d,1,NPOLES);
}
```

fourn is a routine for performing N-dimensional Fourier transforms. We have used it in sample program xfourn.c to transform a 3-dimensional complex data array of dimensions $4 \times 8 \times 16$. The function analyzed is not that easy to visualize, but it is very easy to calculate. The test conducted here is to perform a 3-dimensional transform and inverse transform in succession, and to compare the result with the original array. Ratios are provided for convenience.

```c
/* Driver for routine FOURN */

#include <stdio.h>
#include "nr.h"
#include "nrutil.h"

#define NDIM 3
#define NDAT2 1024

main()
{
    int i,isign,j,k,l,ll,ndum,*nn;
    float *data;

    nn=ivector(1,NDIM);
    data=vector(1,NDAT2);
    ndum=2;
    for (i=1;i<=NDIM;i++) {
        ndum *= 2;
        nn[i]=ndum;
    }
    for (k=1;k<=nn[1];k++)
        for (j=1;j<=nn[2];j++)
            for (i=1;i<=nn[3];i++) {
                l=k+(j-1)*nn[1]+(i-1)*nn[2]*nn[1];
                ll=2*l-1;
                data[ll]=ll;
                data[ll+1]=ll+1;
            }
    isign=1;
    fourn(data,nn,NDIM,isign);
    isign = -1;
    fourn(data,nn,NDIM,isign);
    printf("Double 3-dimensional transform\n\n");
    printf("%22s %24s %20s\n",
        "Double transf.","Original data","Ratio");
    printf("%10s %13s %12s %13s %11s %13s\n\n",
        "real","imag.","real","imag.","real","imag.");
    for (i=1;i<=4;i++) {
        k=2*(j=2*i);
        l=k+(j-1)*nn[1]+(i-1)*nn[2]*nn[1];
        ll=2*l-1;
        printf("%12.2f %12.2f %10d %12d %14.2f %12.2f\n",
            data[ll],data[ll+1],ll,ll+1,data[ll]/ll,
            data[ll+1]/(ll+1));
    }
    printf("\nThe product of transform lengths is: %4d\n",nn[1]*nn[2]*nn[3]);
    free_vector(data,1,NDAT2);
    free_ivector(nn,1,NDIM);
}
```

Chapter 13: Statistical Description of Data

Chapter 13 of *Numerical Recipes* covers the subject of descriptive statistics, the representation of data in terms of its statistical properties, and the use of such properties to compare data sets. There are three procedures that characterize data sets. moment returns the average, average deviation, standard deviation, variance, skewness, and kurtosis of a data array. mdian1 and mdian2 both find the median of an array. The former also sorts the array.

Most of the remaining procedures compare data sets. ttest compares the means of two data sets having the same variance; tutest does the same for two sets having different variance; and tptest does it for paired samples, correcting for covariance. ftest is a test of whether two data arrays have significantly different variance. The question of whether two distributions are different is treated by four procedures (pertaining to whether the data is binned or continuous, and whether data is compared to a model distribution or to other data). Specifically,

1. chsone *compares binned data to a model distribution.*

2. chstwo *compares two binned data sets.*

3. ksone *compares the cumulative distribution function of an unbinned data set to a given function.*

4. kstwo *compares the cumulative distribution functions of two unbinned data sets.*

The next set of procedures tests for associations between nominal variables. cntab1 and cntab2 both check for associations in a two-dimensional contingency table, the first calculating on the basis of χ^2, and the second by evaluating entropies. Linear correlation is represented by Pearson's r, or the linear correlation coefficient, which is calculated with routine pearsn. Alternatively, the data can be investigated with a nonparametric or rank correlation, using spear to find Spearman's rank correlation r_s. Kendall's τ uses rank ordering of ordinal data to test for monotonic correlations. kendl1 *does this for two data arrays of the same size, while* kendl2 *applies it to contingency tables.*

One final routine smoott *makes no attempt to describe or compare data statistically. It seeks, instead, to smooth out the statistical fluctuations, usually for the purpose of visual presentation.*

★ ★ ★ ★

Procedure `moment` calculates successive moments of a given distribution of data. The example program `xmoment.c` creates an unusual distribution, one that has a sinusoidal distribution of values (over a half-period of the sine, so the distribution is a symmetrical peak). We have worked out the moments of such a distribution theoretically and recorded them in the program for comparison. The data is discrete and will only approximate these values.

```c
/* Driver for routine MOMENT */

#include <stdio.h>
#include <math.h>
#include "nr.h"
#include "nrutil.h"

#define PI 3.14159265
#define NPTS 5000
#define NBIN 100
#define NPPNB (NPTS+NBIN)

main()
{
    int i=0,k,nlim;
    float adev,ave,curt,sdev,skew,vrnce,x,*data;

    data=vector(1,NPPNB);
    for (x=PI/NBIN;x<=PI;x += PI/NBIN) {
        nlim=0.5+sin(x)*PI/2.0*NPTS/NBIN;
        for (k=1;k<=nlim;k++) data[++i]=x;
    }
    printf("moments of a sinusoidal distribution\n\n");
    moment(data,i,&ave,&adev,&sdev,&vrnce,&skew,&curt);
    printf("%39s %11s\n\n","calculated","expected");
    printf("%s %17s %12.4f %12.4f\n","Mean :"," ",ave,PI/2.0);
    printf("%s %4s %12.4f %12.4f\n",
        "Average Deviation :"," ",adev,(PI/2.0)-1.0);
    printf("%s %3s %12.4f %12.4f\n",
        "Standard Deviation :"," ",sdev,0.683667);
    printf("%s %13s %12.4f %12.4f\n",
        "Variance :"," ",vrnce,0.467401);
    printf("%s %13s %12.4f %12.4f\n",
        "Skewness :"," ",skew,0.0);
    printf("%s %13s %12.4f %12.4f\n",
        "Kurtosis :"," ",curt,-0.806249);
    free_vector(data,1,NPPNB);
}
```

`mdian1` and `mdian2` both find the median of a distribution. In programs `xm-dian1.c` and `xmdian2.c` we allow this distribution to be Gaussian, as produced by routine `gasdev`. This distribution should have a mean of zero and variance of one. `mdian1` also sorts the data, so `xmdian1.c` prints the sorted data to show that it is done properly. Example `xmdian2.c` has nothing to show from `mdian2` but the median itself, and it is checked by comparing to the result from `mdian1`.

```c
/* Driver for routine MDIAN1 */

#include <stdio.h>
#include "nr.h"
```

```
#include "nrutil.h"

#define NPTS 50

main()
{
    int i,j,idum=(-5);
    float xmed,*data;

    data=vector(1,NPTS);
    for (i=1;i<=NPTS;i++) data[i]=gasdev(&idum);
    mdian1(data,NPTS,&xmed);
    printf("\nData drawn from a gaussian distribution\n");
    printf("with zero mean and unit variance\n\n");
    printf("Median of data set is %9.6f\n\n",xmed);
    printf("Sorted data\n");
    for (i=1;i<=NPTS/5;i++) {
        for (j=1;j<=5;j++) printf("%12.6f",data[5*i-5+j]);
        printf("\n");
    }
    free_vector(data,1,NPTS);
}
/* Driver for routine MDIAN2 */

#include <stdio.h>
#include "nr.h"
#include "nrutil.h"

#define NPTS 50

main()
{
    int i,idum;
    float xmed,*data;

    data=vector(1,NPTS);
    idum=(-5);
    for (i=1;i<=NPTS;i++) data[i]=gasdev(&idum);
    mdian2(data,NPTS,&xmed);
    printf("\nData drawn from a gaussian distribution\n");
    printf("with zero mean, unit variance\n\n");
    printf("median according to mdian2 is %9.6f\n",xmed);
    mdian1(data,NPTS,&xmed);
    printf("median according to mdian1 is %9.6f\n",xmed);
    free_vector(data,1,NPTS);
}
```

Student's t-test is a test of two data sets for significantly different means. It is applied by xttest.c to two Gaussian data sets data1 and data2 that are generated by gasdev. data2 is originally given an artificial shift of its mean to the right of that of data1, by NSHFT/2 units of EPS. Then data1 is successively shifted NSHFT times to the right by EPS and compared to data2 by ttest. At about step NSHFT/2, the two distributions should superpose and indicate populations with the same mean. Notice that the two populations have the same variance (i.e. 1.0), as required by ttest.

```
/* Driver for routine TTEST */

#include <stdio.h>
#include "nr.h"
#include "nrutil.h"

#define NPTS 1024
#define MPTS 512
#define EPS 0.02
#define NSHFT 10

main()
{
    int i,idum,j;
    float prob,t,*data1,*data2;

    data1=vector(1,NPTS);
    data2=vector(1,MPTS);
    /* Generate gaussian distributed data */
    printf("%6s %8s %16s\n","shift","t","probability");
    idum= -5;
    for (i=1;i<=NPTS;i++) data1[i]=gasdev(&idum);
    idum= -11;
    for (i=1;i<=MPTS;i++) data2[i]=(NSHFT/2.0)*EPS+gasdev(&idum);
    for (i=1;i<=NSHFT+1;i++) {
        ttest(data1,NPTS,data2,MPTS,&t,&prob);
        printf("%6.2f %10.2f %10.2f\n",(i-1)*EPS,t,prob);
        for (j=1;j<=NPTS;j++) data1[j] += EPS;
    }
    free_vector(data2,1,MPTS);
    free_vector(data1,1,NPTS);
}
```

avevar is an auxiliary routine for ttest. It finds the average and variance of a data set. The following sample program generates a series of Gaussian distributions for i=1, .., 11, and gives each a shift of $(i-1)$EPS and a variance of i^2. This progression allows you easily to check the operation of avevar "by eye".

```
/* Driver for routine AVEVAR */

#include <stdio.h>
#include "nr.h"
#include "nrutil.h"

#define NPTS 1000
#define EPS 0.1

main()
{
    int i,idum=(-5),j;
    float ave,shift,vrnce,*data;

    data=vector(1,NPTS);
    /* generate gaussian distributed data */
    printf("\n%9s %11s %12s\n","shift","average","variance");
    for (i=1;i<=11;i++) {
        shift=(i-1)*EPS;
```

```
        for (j=1;j<=NPTS;j++)
            data[j]=shift+i*gasdev(&idum);
        avevar(data,NPTS,&ave,&vrnce);
        printf("%8.2f %11.2f %12.2f\n",shift,ave,vrnce);
    }
    free_vector(data,1,NPTS);
}
```

tutest also does Student's *t*-test, but applies to the comparison of means of two
distributions with different variance. The example xtutest.c employs the comparison
used on ttest but gives the two distributions data1 and data2 variances of 1.0 and
4.0 respectively.

```
/* Driver for routine TUTEST */

#include <stdio.h>
#include <math.h>
#include "nr.h"
#include "nrutil.h"

#define NPTS 5000
#define MPTS 1000
#define EPS 0.02
#define VAR1 1.0
#define VAR2 4.0
#define NSHFT 10

main()
{
    int i,idum=(-51773),j;
    float fctr1,fctr2,prob,t,*data1,*data2;

    data1=vector(1,NPTS);
    data2=vector(1,MPTS);
    /* Generate two gaussian distributions of different variance */
    fctr1=sqrt(VAR1);
    for (i=1;i<=NPTS;i++) data1[i]=fctr1*gasdev(&idum);
    fctr2=sqrt(VAR2);
    for (i=1;i<=MPTS;i++)
        data2[i]=NSHFT/2.0*EPS+fctr2*gasdev(&idum);
    printf("\nDistribution #1 : variance = %6.2f\n",VAR1);
    printf("Distribution #2 : variance = %6.2f\n\n",VAR2);
    printf("%7s %8s %16s\n","shift","t","probability");
    for (i=1;i<=NSHFT+1;i++) {
        tutest(data1,NPTS,data2,MPTS,&t,&prob);
        printf("%6.2f %10.2f %11.2f\n",(i-1)*EPS,t,prob);
        for (j=1;j<=NPTS;j++) data1[j] += EPS;
    }
    free_vector(data2,1,MPTS);
    free_vector(data1,1,NPTS);
}
```

tptest goes a step further, and compares two distributions not only having different
variances, but also perhaps having point by point correlations. The example xtptest.c
creates two situations, one with correlated and one with uncorrelated distributions. It does
this by way of three data sets. data1 is a simple Gaussian distribution of zero mean and
unit variance. data2 is data1 plus some additional Gaussian fluctuations of smaller

amplitude. data3 is similar to data2 but generated with independent calls to gasdev so that its fluctuations ought not to have any correlation with those of data1. data1 is then given an offset with respect to the others and they are successively shifted as in previous routines. At each step of the shift tptest was applied. Our results are given below:

	Correlated:			Uncorrelated:	
Shift	T	Probability		T	Probability
.01	2.9264	0.0036		0.6028	0.5469
.02	2.1948	0.0286		0.4521	0.6514
.03	1.4632	0.1440		0.3014	0.7632
.04	0.7316	0.4647		0.1507	0.8802
.05	0.0000	1.0000		0.0000	1.0000
.06	-0.7316	0.4647		-0.1507	0.8802
.07	-1.4632	0.1440		-0.3014	0.7632
.08	-2.1948	0.0286		-0.4521	0.6514
.09	-2.9264	0.0036		-0.6028	0.5469
.10	-3.6580	0.0003		-0.7536	0.4514

```
/* Driver for routine TPTEST */

#include <stdio.h>
#include "nr.h"
#include "nrutil.h"

#define NPTS 500
#define EPS 0.01
#define NSHFT 11
#define ANOISE 0.3

main()
{
    int i,idum=(-5),j;
    float ave1,ave2,ave3,gauss;
    float offset,prob1,prob2,shift,t1,t2;
    float var1,var2,var3,*data1,*data2,*data3;

    data1=vector(1,NPTS);
    data2=vector(1,NPTS);
    data3=vector(1,NPTS);
    printf("%29s %31s\n","Correlated:","Uncorrelated:");
    printf("%7s %11s %17s %11s %17s\n",
        "Shift","t","Probability","t","Probability");
    offset=(NSHFT/2)*EPS;
    for (j=1;j<=NPTS;j++) {
        gauss=gasdev(&idum);
        data1[j]=gauss;
        data2[j]=gauss+ANOISE*gasdev(&idum);
        data3[j]=gasdev(&idum)+ANOISE*gasdev(&idum);
    }
    avevar(data1,NPTS,&ave1,&var1);
    avevar(data2,NPTS,&ave2,&var2);
    avevar(data3,NPTS,&ave3,&var3);
    for (j=1;j<=NPTS;j++) {
        data1[j] -= ave1-offset;
        data2[j] -= ave2;
        data3[j] -= ave3;
```

```
    }
    for (i=1;i<=NSHFT;i++) {
        shift=i*EPS;
        for (j=1;j<=NPTS;j++) {
            data2[j] += EPS;
            data3[j] += EPS;
        }
        tptest(data1,data2,NPTS,&t1,&prob1);
        tptest(data1,data3,NPTS,&t2,&prob2);
        printf("%6.2f %14.4f %12.4f %16.4f %12.4f\n",
            shift,t1,prob1,t2,prob2);
    }
    free_vector(data3,1,NPTS);
    free_vector(data2,1,NPTS);
    free_vector(data1,1,NPTS);
}
```

The *F*-test (procedure `ftest`) is a test for differing variances between two distributions. For demonstration purposes, sample program `xftest.c` generates a Gaussian distribution `data1` of unit variance. The values of a second Gaussian distribution `data2` are then set by multiplying `data1` by a series of values `factr` which takes its variance from 1.0 to 1.4 in ten equal steps. The effect of this on the *F*-test can be evaluated from the probabilities `prob`.

```
/* Driver for routine FTEST */

#include <stdio.h>
#include <math.h>
#include "nr.h"
#include "nrutil.h"

#define NPTS 1000
#define MPTS 500
#define EPS 0.01
#define NVAL 11

main()
{
    int i,idum=(-13),j;
    float f,factor,prob,vrnce,*data1,*data2,*data3;

    data1=vector(1,NPTS);
    data2=vector(1,MPTS);
    data3=vector(1,MPTS);
    /* Generate two gaussian distributions with
    different variances */
    printf("\n%16s %5.2f\n","Variance 1 = ",1.0);
    printf("%13s %11s %16s\n","Variance 2","Ratio","Probability");
    for (j=1;j<=NPTS;j++) data1[j]=gasdev(&idum);
    for (j=1;j<=MPTS;j++) data2[j]=gasdev(&idum);
    for (i=1;i<=NVAL;i++) {
        vrnce=1.0+(i-1)*EPS;
        factor=sqrt(vrnce);
        for (j=1;j<=MPTS;j++) data3[j]=factor*data2[j];
        ftest(data1,NPTS,data3,MPTS,&f,&prob);
        printf("%11.4f %13.4f %13.4f\n",vrnce,f,prob);
    }
```

```
        free_vector(data3,1,MPTS);
        free_vector(data2,1,MPTS);
        free_vector(data1,1,NPTS);
}
```

chsone and chstwo compare two distributions on the basis of a χ^2 test to see if they are different. chsone, specifically, compares a data distribution to an expected distribution. xchsone.c generates an exponential distribution bins[i] of data using routine expdev. It then creates an array ebins[i] which is the expected result (a smooth exponential decay in the absence of statistical fluctuations). ebins and bins are compared by chsone to give χ^2 and a probability that they represent the same distribution.

```
/* Driver for routine CHSONE */

#include <stdio.h>
#include <math.h>
#include "nr.h"
#include "nrutil.h"

#define NBINS 10
#define NPTS 2000

main()
{
    int i,ibin,idum=(-15),j;
    float chsq,df,prob,x,*bins,*ebins;

    bins=vector(1,NBINS);
    ebins=vector(1,NBINS);
    for (j=1;j<=NBINS;j++) bins[j]=0.0;
    for (i=1;i<=NPTS;i++) {
        x=expdev(&idum);
        ibin=(int) (x*NBINS/3.0)+1;
        if (ibin <= NBINS) ++bins[ibin];
    }
    for (i=1;i<=NBINS;i++)
        ebins[i]=3.0*NPTS/NBINS*exp(-3.0*(i-0.5)/NBINS);
    chsone(bins,ebins,NBINS,-1,&df,&chsq,&prob);
    printf("%15s %15s\n","expected","observed");
    for (i=1;i<=NBINS;i++)
        printf("%14.2f %15.2f\n",ebins[i],bins[i]);
    printf("\n%19s %10.4f\n","chi-squared:",chsq);
    printf("%19s %10.4f\n","probability:",prob);
    free_vector(ebins,1,NBINS);
    free_vector(bins,1,NBINS);
}
```

chstwo compares two binned distributions bins1 and bins2, again using a χ^2 test. Sample program xchstwo.c prepares these distributions both in the same way. Each is composed of 2000 random numbers, drawn from an exponential deviate, and placed into 10 bins. The two data sets are then analyzed by chstwo to calculate χ^2 and probability prob.

```
/* Driver for routine CHSTWO */

#include <stdio.h>
```

```
#include <math.h>
#include "nr.h"
#include "nrutil.h"

#define NBINS 10
#define NPTS 2000

main()
{
    int i,ibin,idum=(-17),j;
    float chsq,df,prob,x,*bins1,*bins2;

    bins1=vector(1,NBINS);
    bins2=vector(1,NBINS);
    for (j=1;j<=NBINS;j++) {
        bins1[j]=0.0;
        bins2[j]=0.0;
    }
    for (i=1;i<=NPTS;i++) {
        x=expdev(&idum);
        ibin=x*NBINS/3.0+1;
        if (ibin <= NBINS) ++bins1[ibin];
        x=expdev(&idum);
        ibin=x*NBINS/3.0+1;
        if (ibin <= NBINS) ++bins2[ibin];
    }
    chstwo(bins1,bins2,NBINS,-1,&df,&chsq,&prob);
    printf("\n%15s %15s\n","dataset 1","dataset 2");
    for (i=1;i<=NBINS;i++)
        printf("%13.2f %15.2f\n",bins1[i],bins2[i]);
    printf("\n%18s %12.4f\n","chi-squared:",chsq);
    printf("%18s %12.4f\n","probability:",prob);
    free_vector(bins2,1,NBINS);
    free_vector(bins1,1,NBINS);
}
```

The Kolmogorov-Smirnov test used in `ksone` and `kstwo` applies to unbinned distributions with a single independent variable. `ksone` uses the K-S criterion to compare a single data set to an expected distribution, and `kstwo` uses it to compare two data sets. Sample program `xksone.c` creates data sets with Gaussian distributions and with stepwise increasing variance, and compares their cumulative distribution function to the expected result for a Gaussian distribution of unit variance. This result is the error function and is generated by routine `erf`. Increasing variance in the test destribution should reduce the likelihood that it was drawn from the same distribution represented by the comparison function.

```
/* Driver for routine KSONE */

#include <stdio.h>
#include <math.h>
#include "nr.h"
#include "nrutil.h"

#define NPTS 1000
#define EPS 0.1
```

```
float func(x)
float x;
{
    return 1.0 - erfcc(x/sqrt(2.0));
}

main()
{
    int i,idum=(-5),j;
    float d,factr,prob,varnce,*data;

    data=vector(1,NPTS);
    printf("%19s %16s %15s\n\n",
        "variance ratio","k-s statistic","probability");
    for (i=1;i<=11;i++) {
        varnce=1.0+(i-1)*EPS;
        factr=sqrt(varnce);
        for (j=1;j<=NPTS;j++)
            data[j]=factr*fabs(gasdev(&idum));
        ksone(data,NPTS,func,&d,&prob);
        printf("%16.6f %16.6f %16.6f \n",varnce,d,prob);
    }
    free_vector(data,1,NPTS);
}
```

kstwo compares the cumulative distribution functions of two unbinned data sets, data1 and data2. In sample program xkstwo.c, they are both Gaussian distributions, but data2 is given a stepwise increase of variance. In other respects, xkstwo.c is like xksone.c.

```
/* Driver for routine KSTWO */

#include <stdio.h>
#include <math.h>
#include "nr.h"
#include "nrutil.h"

#define N1 2000
#define N2 1000
#define EPS 0.1

main()
{
    int i,idum=(-1357),j;
    float d,factr,prob,varnce,*data1,*data2;

    data1=vector(1,N1);
    data2=vector(1,N2);
    for (i=1;i<=N1;i++) data1[i]=gasdev(&idum);
    printf("%18s %15s %14s\n",
        "variance ratio","k-s statistic","probability");
    idum = -2468;
    for (i=1;i<=11;i++) {
        varnce=1.0+(i-1)*EPS;
        factr=sqrt(varnce);
        for (j=1;j<=N2;j++)
            data2[j]=factr*gasdev(&idum);
```

```
        kstwo(data1,N1,data2,N2,&d,&prob);
        printf("%15.6f %15.6f %15.6f\n",varnce,d,prob);
    }
    free_vector(data2,1,N2);
    free_vector(data1,1,N1);
}
```

probks is an auxiliary routine for ksone and kstwo which calculates the function $Q_{ks}(\lambda)$ used to evaluate the probability that the two distributions being compared are the same. There is no independent means of producing this function, so in sample program xprobks.c we have chosen simply to graph it. Our output is reproduced below.

```
/* Driver for routine PROBKS */

#include <stdio.h>
#include "nr.h"

#define NPTS 20
#define EPS 0.1
#define ISCAL 40

main()
{
    int i,j,jmax;
    char txt[ISCAL+1];
    float alam,aval;

    printf("probability function for kolmogorov-smirnov statistic\n\n");
    printf("%7s %10s %13s\n","lambda","value:","graph:");
    for (i=1;i<=NPTS;i++) {
        alam=i*EPS;
        aval=probks(alam);
        jmax=(int) (0.5+(ISCAL-1)*aval);
        for (j=0;j<=ISCAL-1;j++) {
            if (j <= jmax)
                txt[j]='*';
            else
                txt[j]=' ';
        }
        txt[ISCAL]='\0';
        printf("%8.6f %10.6f        %s\n",alam,aval,txt);
    }
}
```

```
Probability func. for Kolmogorov-Smirnov statistic
 Lambda:     Value:      Graph:
 0.100000    1.000000    ****************************************
 0.200000    1.000000    ****************************************
 0.300000    0.999991    ****************************************
 0.400000    0.997192    ****************************************
 0.500000    0.963945    ****************************************
 0.600000    0.864283    **********************************
 0.700000    0.711235    ****************************
 0.800000    0.544142    *********************
 0.900000    0.392731    ****************
 1.000000    0.270000    ***********
 1.100000    0.177718    *******
 1.200000    0.112250    ****
```

```
1.300000    0.068092    ***
1.400000    0.039682    **
1.500000    0.022218    *
1.600000    0.011952
1.700000    0.006177
1.800000    0.003068
1.900000    0.001464
2.000000    0.000671
```

Procedure cntab1 analyzes a two-dimensional contingency table and returns several parameters describing any association between its nominal variables. Sample program xcntab1.c supplies a table from a file table1.dat which is listed in the Appendix to this chapter. The table shows the rate of certain accidents, tabulated on a monthly basis. These data are listed, as well as their statistical properties, by the program. We found the results to be:

```
Chi-squared                 5026.30
Degrees of Freedom            88.00
Probability                    .0000
Cramer-V                       .0772
Contingency Coeff.             .2134
```

```c
/* Driver for routine CNTAB1 */

#include <stdio.h>
#include "nr.h"
#include "nrutil.h"

#define NDAT 9
#define NMON 12
#define MAXSTR 80

main()
{
    int i,j,**nmbr;
    float ccc,chisq,cramrv,df,prob;
    char dummy[MAXSTR],fate[NDAT+1][16],mon[NMON+1][6],txt[16];
    FILE *fp;

    nmbr=imatrix(1,NDAT,1,NMON);
    if ((fp = fopen("table1.dat","r")) == NULL)
        nrerror("Data file TABLE1.DAT not found\n");
    fgets(dummy,MAXSTR,fp);
    fgets(dummy,MAXSTR,fp);
    fscanf(fp,"%16c",txt);
    txt[15]='\0';
    for (i=1;i<=12;i++) fscanf(fp," %s",mon[i]);
    fgets(dummy,MAXSTR,fp);
    fgets(dummy,MAXSTR,fp);
    for (i=1;i<=NDAT;i++) {
        fscanf(fp,"%16[^0123456789]",fate[i]);
        fate[i][15]='\0';
        for (j=1;j<=12;j++)
            fscanf(fp,"%d ",&nmbr[i][j]);
    }
    fclose(fp);
```

```
        printf("\n%s",txt);
        for (i=1;i<=12;i++) printf("%5s",mon[i]);
        printf("\n\n");
        for (i=1;i<=NDAT;i++) {
            printf("%s",fate[i]);
            for (j=1;j<=12;j++) printf("%5d",nmbr[i][j]);
            printf("\n");
        }
        cntab1(nmbr,NDAT,NMON,&chisq,&df,&prob,&cramrv,&ccc);
        printf("\n%15s chi-squared        %20.2f\n"," ",chisq);
        printf("%15s degrees of freedom%20.2f\n"," ",df);
        printf("%15s probability       %20.4f\n"," ",prob);
        printf("%15s cramer-v          %20.4f\n"," ",cramrv);
        printf("%15s contingency coeff.%20.4f\n"," ",ccc);
        free_imatrix(nmbr,1,NDAT,1,NMON);
}
```

The test looks for any association between accidents and the months in which they occur. `table1.dat` clearly shows some. Drownings, for example, happen mostly in the summer. `cntab2` carries out a similar analysis on `table1.dat` but measures associations on the basis of entropy. Sample program `xcntab2.c` prints out the following entropies for the table:

```
Entropy of Table              4.0368
Entropy of x-distribution     1.5781
Entropy of y-distribution     2.4820
Entropy of y given x          2.4588
Entropy of x given y          1.5548
Dependency of y on x           .0094
Dependency of x on y           .0147
Symmetrical dependency         .0114
```

```
/* Driver for routine CNTAB2 */

#include <stdio.h>
#include "nr.h"
#include "nrutil.h"

#define NI 9
#define NMON 12
#define MAXSTR 80

main()
{
    float h,hx,hxgy,hy,hygx,uxgy,uxy,uygx;
    int i,j,**nmbr;
    char dummy[MAXSTR],fate[NI+1][16],mon[NMON+1][6],txt[16];
    FILE *fp;

    nmbr=imatrix(1,NI,1,NMON);
    if ((fp = fopen("table1.dat","r")) == NULL)
        nrerror("Data file TABLE1.DAT not found\n");
    fgets(dummy,MAXSTR,fp);
    fgets(dummy,MAXSTR,fp);
    fscanf(fp,"%16c",txt);
    txt[15]='\0';
    for (i=1;i<=12;i++) fscanf(fp," %s",mon[i]);
```

```
        fgets(dummy,MAXSTR,fp);
        fgets(dummy,MAXSTR,fp);
        for (i=1;i<=NI;i++) {
            fscanf(fp,"%16[^0123456789]",fate[i]);
            fate[i][15]='\0';
            for (j=1;j<=12;j++)
                fscanf(fp,"%d ",&nmbr[i][j]);
        }
        fclose(fp);
        printf("\n%s",txt);
        for (i=1;i<=12;i++) printf("%5s",mon[i]);
        printf("\n\n");
        for (i=1;i<=NI;i++) {
            printf("%s",fate[i]);
            for (j=1;j<=12;j++) printf("%5d",nmbr[i][j]);
            printf("\n");
        }
        cntab2(nmbr,NI,NMON,&h,&hx,&hy,&hygx,&hxgy,&uygx,&uxgy,&uxy);
        printf("\n          entropy of table             %10.4f\n",h);
        printf("         entropy of x-distribution   %10.4f\n",hx);
        printf("         entropy of y-distribution   %10.4f\n",hy);
        printf("         entropy of y given x        %10.4f\n",hygx);
        printf("         entropy of x given y        %10.4f\n",hxgy);
        printf("         dependency of y on x        %10.4f\n",uygx);
        printf("         dependency of x on y        %10.4f\n",uxgy);
        printf("         symmetrical dependency      %10.4f\n",uxy);
        free_imatrix(nmbr,1,NI,1,NMON);
}
```

The dependencies of x on y and y on x indicate the degree to which the type of accident can be predicted by knowing the month, or vice-versa.

pearsn makes an examination of two ordinal or continuous variables to find linear correlations. It returns a linear correlation coefficient r, a probability of correlation prob, and Fisher's z. Sample program xpearsn.c sets up data pairs in arrays dose and spore which show hypothetical data for the spore count from plants exposed to various levels of γ-rays. The results of applying pearsn to this data set are compared with the correct results by the program.

```
/* Driver for routine PEARSN */

#include <stdio.h>
#include "nr.h"

#define N 10

main()
{
    int i;
    float prob,r,z;
    static float dose[N+1]=
        {0.0,56.1,64.1,70.0,66.6,82.0,91.3,90.0,99.7,115.3,110.0};
    static float spore[N+1]=
        {0.0,0.11,0.40,0.37,0.48,0.75,0.66,0.71,1.20,1.01,0.95};

    printf("\nEffect of Gamma Rays on Man-in-the-Moon Marigolds\n");
    printf("%16s %23s\n","Count Rate (cpm)","Pollen Index");
```

```
    for (i=1;i<=N;i++)
        printf("%10.2f %25.2f \n",dose[i],spore[i]);
    pearsn(dose,spore,N,&r,&prob,&z);
    printf("\n%30s %16s\n","PEARSN","Expected");
    printf("%s %8s %9f %15f\n","Corr. Coeff."," ",r,0.9069586);
    printf("%s %9s %9f %15f\n","Probability"," ",prob,0.2926505e-3);
    printf("%s %10s %9f %15f\n","Fisher's z"," ",z,1.510110);
}
```

Rank order correlation may be done with spear to compare two distributions data1 and data2 for correlation. Correlations are reported both in terms of d, the sum-squared difference in ranks, and rs, Spearman's rank correlation parameter. Sample program xspear.c applies the calculation to the data in table table2.dat (see Appendix) which shows the solar flux incident on various cities during different months of the year. It then checks for correlations between columns of the table, considering each column as a separate data set. In this fashion it looks for correlations between the July solar flux and that of other months. The probability of such correlations are shown by probd and probrs. Our results are:

```
Correlation of sampled U.S. solar radiation (July with other months)
Month        D        St. Dev.      PROBD      Spearman R     PROBRS
  jul       .00      -4.358899     .000013      .993965      .000000
  aug    122.00      -3.958458     .000075      .901959      .000000
  sep    218.00      -3.643896     .000269      .832831      .000005
  oct    384.00      -3.098495     .001945      .704372      .000526
  nov    390.50      -3.077642     .002086      .701205      .000572
  dec    622.00      -2.318075     .020445      .526751      .017022
  jan    644.50      -2.244251     .024816      .509796      .021662
  feb    483.50      -2.772503     .005563      .631122      .002844
  mar    497.00      -2.728208     .006368      .620949      .003480
  apr    405.50      -3.027925     .002462      .688158      .000796
  may    264.00      -3.492371     .000479      .794870      .000028
  jun    121.50      -3.960099     .000075      .902336      .000000
```

```
/* Driver for routine SPEAR */

#include <stdio.h>
#include "nr.h"
#include "nrutil.h"

#define NDAT 20
#define NMON 12
#define MAXSTR 80

main()
{
    int i,j;
    float d,probd,probrs,rs,zd,*ave,*data1,*data2,*zlat,**rays;
    char dummy[MAXSTR],txt[MAXSTR],city[NDAT+1][17],mon[NMON+1][5];
    FILE *fp;

    ave=vector(1,NDAT);
    data1=vector(1,NDAT);
    data2=vector(1,NDAT);
    zlat=vector(1,NDAT);
    rays=matrix(1,NDAT,1,NMON);
```

```
    if ((fp = fopen("table2.dat","r")) == NULL)
        nrerror("Data file TABLE2.DAT not found\n");
    fgets(dummy,MAXSTR,fp);
    fgets(txt,MAXSTR,fp);
    fscanf(fp,"%*15c");
    for (i=1;i<=NMON;i++) fscanf(fp," %s",mon[i]);
    fgets(dummy,MAXSTR,fp);
    fgets(dummy,MAXSTR,fp);
    for (i=1;i<=NDAT;i++) {
        fscanf(fp,"%[^0123456789]",city[i]);
        city[i][16]='\0';
        for (j=1;j<=NMON;j++) fscanf(fp,"%f",&rays[i][j]);
        fscanf(fp,"%f %f ",&ave[i],&zlat[i]);
    }
    fclose(fp);
    printf("%s\n",txt);
    printf("%16s"," ");
    for (i=1;i<=12;i++) printf("%4s",mon[i]);
    printf("\n");
    for (i=1;i<=NDAT;i++) {
        printf("%s",city[i]);
        for (j=1;j<=12;j++)
            printf("%4d",(int) (0.5+rays[i][j]));
        printf("\n");
    }
    /* Check temperature correlations between different months */
    printf("\nAre sunny summer places also sunny winter places?\n");
    printf("Check correlation of sampled U.S. solar radiation\n");
    printf("(july with other months)\n\n");
    printf("%s %9s %14s %11s %15s %10s\n","month","d",
        "st. dev.","probd","spearman-r","probrs");
    for (i=1;i<=NDAT;i++) data1[i]=rays[i][1];
    for (j=1;j<=12;j++) {
        for (i=1;i<=NDAT;i++) data2[i]=rays[i][j];
        spear(data1,data2,NDAT,&d,&zd,&probd,&rs,&probrs);
        printf("%4s %12.2f %12.6f %12.6f %13.6f %12.6f\n",
            mon[j],d,zd,probd,rs,probrs);
    }
    free_matrix(rays,1,NDAT,1,NMON);
    free_vector(zlat,1,NDAT);
    free_vector(data2,1,NDAT);
    free_vector(data1,1,NDAT);
    free_vector(ave,1,NDAT);
}
```

crank is an auxiliary routine for spear and is used in conjunction with sort2.
The latter sorts an array, and crank then assigns ranks to each data entry, including the
midranking of ties. Sample program xcrank.c uses the solar flux data of table2.dat
(see Appendix) to illustrate. Each column of the solar flux table is replaced by the rank
order of its entries. You can check the rank order chart against the chart of original
values to verify the ordering.

```
/* Driver for routine CRANK */

#include <stdio.h>
#include "nr.h"
#include "nrutil.h"
```

```
#define NDAT 20
#define NMON 12
#define MAXSTR 80

main()
{
    int i,j;
    float *data,*order,*s,**rays;
    char dummy[MAXSTR],txt[MAXSTR],city[NDAT+1][17],mon[NMON+1][5];
    FILE *fp;

    data=vector(1,NDAT);
    order=vector(1,NDAT);
    s=vector(1,NMON);
    rays=matrix(1,NDAT,1,NMON);
    if ((fp = fopen("table2.dat","r")) == NULL)
        nrerror("Data file TABLE2.DAT not found\n");
    fgets(dummy,MAXSTR,fp);
    fgets(txt,MAXSTR,fp);
    fscanf(fp,"%*15c");
    for (i=1;i<=NMON;i++) fscanf(fp," %s",mon[i]);
    fgets(dummy,MAXSTR,fp);
    fgets(dummy,MAXSTR,fp);
    for (i=1;i<=NDAT;i++) {
        fscanf(fp,"%[^0123456789]",city[i]);
        city[i][16]='\0';
        for (j=1;j<=NMON;j++) fscanf(fp,"%f",&rays[i][j]);
        fgets(dummy,MAXSTR,fp);
    }
    fclose(fp);
    printf("%s\n%15s",txt," ");
    for (i=1;i<=12;i++) printf(" %s",mon[i]);
    printf("\n");
    for (i=1;i<=NDAT;i++) {
        printf("%s",city[i]);
        for (j=1;j<=12;j++)
            printf("%4d",(int) (0.5+rays[i][j]));
        printf("\n");
    }
    printf(" press return to continue ...\n");
    getchar();
    /* Replace solar flux in each column by rank order */
    for (j=1;j<=12;j++) {
        for (i=1;i<=NDAT;i++) {
            data[i]=rays[i][j];
            order[i]=i;
        }
        sort2(NDAT,data,order);
        crank(NDAT,data,&s[j]);
        for (i=1;i<=NDAT;i++)
            rays[(int) (0.5+order[i])][j]=data[i];
    }
    printf("%15s"," ");
    for (i=1;i<=12;i++) printf(" %s",mon[i]);
    printf("\n");
    for (i=1;i<=NDAT;i++) {
```

```
            printf("%s",city[i]);
            for (j=1;j<=12;j++)
                printf("%4d",(int) (0.5+rays[i][j]));
            printf("\n");
        }
        free_matrix(rays,1,NDAT,1,NMON);
        free_vector(s,1,NMON);
        free_vector(order,1,NDAT);
        free_vector(data,1,NDAT);
}
```

kendl1 and kendl2 test for monotonic correlations of ordinal data. They differ
in that kendl1 compares two data sets of the same rank, while kendl2 operates on a
contingency table. Sample program xkendl1.c, for example, looks for pair correlations
in three of our random number routines. That is to say, it tests for randomness by seeing
if two consecutive numbers from the generator have a monotonic correlation. It uses the
random number generators ran0, ran3, and ran4, one at a time, to generate 200 pairs
of random numbers each. Then kendl1 tests for correlation of the pairs, and a chart
is made showing Kendall's τ, the standard deviation from the null hypotheses, and the
probability. For a better test of the generators, you may wish to increase the number of
pairs NDAT. It would also be a good idea to see how your result depends on the value
of the seed idum.

```c
/* Driver for routine KENDL1 */

#include <stdio.h>
#include "nr.h"
#include "nrutil.h"

#define NDAT 200

main()
{
    int i,idum,j;
    float prob,tau,z,*data1,*data2;
    static char *txt[3]={"RAN0","RAN3","RAN4"};

    data1=vector(1,NDAT);
    data2=vector(1,NDAT);
    /* Look for correlations in RAN0, RAN3 and RAN4 */
    printf("\nPair correlations of RAN0, RAN3 and RAN4\n\n");
    printf("%9s %17s %16s %18s\n",
        "Program","Kendall tau","Std. Dev.","Probability");
    for (i=1;i<=3;i++) {
        idum=(-1357);
        for (j=1;j<=NDAT;j++) {
            if (i == 1) {
                data1[j]=ran0(&idum);
                data2[j]=ran0(&idum);
            } else if (i == 2) {
                data1[j]=ran3(&idum);
                data2[j]=ran3(&idum);
            } else if (i == 3) {
                data1[j]=ran4(&idum);
                data2[j]=ran4(&idum);
            }
```

```
        }
        kendl1(data1,data2,NDAT,&tau,&z,&prob);
        printf("%8s %17.6f %17.6f %17.6f\n",txt[i-1],tau,z,prob);
    }
    free_vector(data2,1,NDAT);
    free_vector(data1,1,NDAT);
}
```

Sample program xkendl2.c prepares a contingency table based on the routines irbit1 and irbit2. You may recall that these routines generate random binary sequences. The program checks the sequences by breaking them into groups of three bits. Each group is treated as a three-bit binary number. Two consecutive groups then act as indices into an 8×8 contingency table that records how many times each possible sequence of six bits (two groups) occurs. For each random bit generator, NDAT=1000 samples are taken. Then the contingency table tab[k][l] is analyzed by kendl2 to find Kendall's τ, the standard deviation, and the probability. Notice that Kendall's τ can only be applied when both variables are ordinal (here, the numbers 0 to 7), and that the test is specifically for monotonic correlations. In this case we are actually testing whether the larger 3-bit binary numbers tend to be followed by others of their own kind. Within the program, we have expressed this roughly as a test of whether ones or zeros tend to come in groups more than they should.

```
/* Driver for routine KENDL2 */

#include <stdio.h>
#include "nr.h"
#include "nrutil.h"

#define NDAT 1000
#define IP 8
#define JP 8

main()
{
    int ifunc,i=IP,j=JP,k,l,m,n,twoton;
    unsigned long int iseed;
    float prob,tau,z,**tab;
    static char *txt[8]=
        {"000","001","010","011","100","101","110","111"};

    /* Look for 'ones-after-zeros' in IRBIT1 and IRBIT2 sequences */
    tab=matrix(1,IP,1,JP);
    printf("Are ones followed by zeros and vice-versa?\n");
    for (ifunc=1;ifunc<=2;ifunc++) {
        iseed=2468;
        if (ifunc == 1)
            printf("test of irbit1:\n");
        else
            printf("test of irbit2:\n");
        for (k=1;k<=i;k++)
            for (l=1;l<=j;l++) tab[k][l]=0.0;
        for (m=1;m<=NDAT;m++) {
            k=1;
            twoton=1;
            for (n=0;n<=2;n++) {
                if (ifunc == 1)
```

```
                    k += (irbit1(&iseed)*twoton);
                else
                    k += (irbit2(&iseed)*twoton);
                twoton *= 2;
            }
            l=1;
            twoton=1;
            for (n=0;n<=2;n++) {
                if (ifunc == 1)
                    l += (irbit1(&iseed)*twoton);
                else
                    l += (irbit2(&iseed)*twoton);
                twoton *= 2;
            }
            tab[k][l] += 1.0;
        }
        kendl2(tab,i,j,&tau,&z,&prob);
        printf("      ");
        for (n=0;n<=7;n++) printf("%6s",txt[n]);
        printf("\n");
        for (n=1;n<=8;n++) {
            printf("%3s",txt[n-1]);
            for (m=1;m<=8;m++)
                printf("%6d",(int) (0.5+tab[n][m]));
            printf("\n");
        }
        printf("\n%17s %14s %16s\n",
            "kendall tau","std. dev.","probability");
        printf("%15.6f %15.6f %15.6f\n\n",tau,z,prob);
    }
    free_matrix(tab,1,IP,1,JP);
}
```

smooft is a procedure for smoothing data. This is not a mathematically valuable procedure since it always reduces the information content of the data. However, it is a satisfactory tool for data presentation, as it may help to make evident important features of the data. Sample program xsmooft.c prepares an artificial data set y[i] with a broad maximum. It then adds noise from a Gaussian deviate. Subsequently the data is plotted three times; first the original data, then following each of two consecutive applications of smooft. You will notice that the second use of smooft is almost entirely ineffectual, but the first makes a significant change in the presentational quality of the graph.

```
/* Driver for routine SMOOFT */

#include <stdio.h>
#include <math.h>
#include "nr.h"
#include "nrutil.h"

#define N 100
#define HASH 0.05
#define SCALE 100.0
#define PTS 10.0
#define M 256          /* first integral power of 2 that
            is greater or equal to (N+2*PTS) */

main()
```

```
{
    int i,idum=(-7),j,k,nstp,bar;
    float *y;
    char txt[52];

    y=vector(1,M);
    for (i=1;i<=N;i++) {
        y[i]=3.0*i/N*exp(-3.0*i/N);
        y[i] += (HASH*gasdev(&idum));
    }
    for (k=1;k<=3;k++) {
        nstp=N/20;
        printf("\n%8s %12s\n","data","graph");
        for (i=1;i<=20*nstp;i += nstp) {
            bar=(int) (0.5+SCALE*y[i]);
            for (j=1;j<=50;j++)
                txt[j]=(j <= bar ? '*': ' ');
            txt[51]='\0';
            printf("%10.6f %4s %s\n",y[i]," ",txt+1);
        }
        if (k == 3) break;
        printf("press return to smooth ...\n");
        getchar();
        smooft(y,N,PTS);
    }
    free_vector(y,1,M);
}
```

Appendix

File table1.dat:

Accidental Deaths by Month and Type (1979)

Month:	jan	feb	mar	apr	may	jun	jul	aug	sep	oct	nov	dec
Motor Vehicle	3298	3304	4241	4291	4594	4710	4914	4942	4861	4914	4563	4892
Falls	1150	1034	1089	1126	1142	1100	1112	1099	1114	1079	999	1181
Drowning	180	190	370	530	800	1130	1320	990	580	320	250	212
Fires	874	768	630	516	385	324	277	272	271	381	533	760
Choking	299	264	258	247	273	269	251	269	271	279	297	266
Fire-arms	168	142	122	140	153	142	147	160	162	172	266	230
Poisons	298	277	346	263	253	239	268	228	240	260	252	241
Gas-poison	267	193	144	127	70	63	55	53	60	118	150	172
Other	1264	1234	1172	1220	1547	1339	1419	1453	1359	1308	1264	1246

File table2.dat:

Average solar radiation (watts/square meter) for selected cities

Month:	jul	aug	sep	oct	nov	dec	jan	feb	mar	apr	may	jun	ave	lat
Atlanta, GA	257	246	201	166	30	102	106	140	184	236	258	271	192	34.0
Barrow, AK	208	123	56	20	0	0	0	18	87	184	248	256	100	71.0
Bismark, ND	296	251	185	132	78	60	76	121	170	217	267	284	178	47.0
Boise, ID	324	275	221	152	88	60	69	113	164	235	284	309	191	43.5
Boston, MA	240	206	165	115	70	58	67	96	142	176	228	242	150	42.5
Caribou, ME	246	218	161	102	53	51	66	111	178	194	229	232	153	47.0
Cleveland, OH	267	239	182	127	68	56	60	87	151	182	253	271	162	41.5
Dodge City, KS	311	287	239	184	138	113	123	153	202	256	275	315	216	38.0
El Paso, TX	324	309	278	224	178	151	160	209	266	317	346	353	260	32.0
Fresno, CA	323	293	243	182	117	77	90	143	212	264	308	337	216	37.0

Greensboro, NC	263	235	197	156	118	95	97	134	171	227	257	273	185	36.0
Honolulu, HI	305	293	271	245	208	176	175	200	234	262	300	297	247	21.0
Little Rock, AR	270	250	214	167	118	91	96	127	173	220	256	272	188	35.0
Miami, FL	260	246	216	188	171	154	166	201	238	263	267	257	219	26.0
New York, NY	251	238	175	127	77	62	71	102	151	183	220	255	159	41.0
Omaha, NE	275	252	192	142	96	80	99	134	172	224	248	272	182	21.0
Rapid City, SD	288	262	208	152	99	76	90	135	193	235	259	287	190	44.0
Seattle, WA	242	209	150	84	44	29	34	60	118	174	216	228	132	47.5
Tucson, AZ	304	286	281	216	172	144	151	195	264	322	358	343	253	41.0
Washington, DC	267	190	196	145	75	64	101	124	153	182	215	247	163	39.0

Chapter 14: Modeling of Data

Chapter 14 of Numerical Recipes deals with the fitting of a model function to a set of data, in order to summarize the data in terms of a few model parameters. Both traditional least-squares fitting and robust fitting are considered. Fits to a straight line are carried out by routine fit. *More general linear least-squares fits are handled by* lfit *and* covsrt. *(Remember that the term "linear" here refers not to a linear dependence of the fitting function on its argument, but rather to a linear dependence of the function on its fitting parameters.) In cases where* lfit *fails, owing probably to near degeneracy of some basis functions, the answer may still be found using* svdfit *and* svdvar. *In fact, these are generally recommended in preference to* lfit *because they never (?) fail. For nonlinear least-squares fits, the Levenberg-Marquardt method is discussed, and is implemented in* mrqmin, *which makes use also of* covsrt *and* mrqcof.

Robust estimation is discussed in several forms, and illustrated by routine medfit *which fits a straight line to data points based on the criterion of least absolute deviations rather than least-squared deviations.* rofunc *is an auxiliary function for* medfit.

★ ★ ★ ★

Routine fit fits a set of N data points (x[i],y[i]), with standard deviations sig[i], to the linear model $y = Ax + B$. It uses χ^2 as the criterion for goodness-of-fit. To demonstrate fit, we generate some noisy data in sample program xfit.c. For NPT values of i we take $x = 0.1i$ and $y = -2x + 1$ plus some values drawn from a Gaussian distribution to represent noise. Then we make two calls to fit, first performing the fit without allowance for standard deviations sig[i], and then with such allowance. Since sig[i] has been set to the constant value SPREAD, it should not affect the resulting parameter values. The values output from this routine are:

Ignoring standard deviation:

```
A =    .936991      Uncertainty:   .099560
B = -1.979427      Uncertainty:   .017116
Chi-squared:      23.922630
Goodness-of-fit:   1.000000
```

Including standard deviation:

```
A =    .936991      Uncertainty:   .100755
B = -1.979427      Uncertainty:   .017321
Chi-squared:      95.690510
Goodness-of-fit:    .547181
```

```
/* Driver for routine FIT */

#include <stdio.h>
#include "nr.h"
#include "nrutil.h"

#define NPT 100
#define SPREAD 0.5

main()
{
    int i,idum=(-117),mwt;
    float a,b,chi2,q,siga,sigb,*x,*y,*sig;

    x=vector(1,NPT);
    y=vector(1,NPT);
    sig=vector(1,NPT);
    for (i=1;i<=NPT;i++) {
        x[i]=0.1*i;
        y[i] = -2.0*x[i]+1.0+SPREAD*gasdev(&idum);
        sig[i]=SPREAD;
    }
    for (mwt=0;mwt<=1;mwt++) {
        fit(x,y,NPT,sig,mwt,&a,&b,&siga,&sigb,&chi2,&q);
        if (mwt == 0)
            printf("\nIgnoring standard deviations\n");
        else
            printf("\nIncluding standard deviations\n");
        printf("%12s %9.6f %18s %9.6f \n",
            "a = ",a,"uncertainty:",siga);
        printf("%12s %9.6f %18s %9.6f \n",
            "b = ",b,"uncertainty:",sigb);
        printf("%19s %14.6f \n","chi-squared: ",chi2);
        printf("%23s %10.6f \n","goodness-of-fit: ",q);
    }
    free_vector(sig,1,NPT);
    free_vector(y,1,NPT);
    free_vector(x,1,NPT);
}
```

lfit carries out the same sort of fit but this time does a linear least-squares fit to a more general function. In sample program xlfit.c the chosen function is a linear sum of powers of x, generated by procedure funcs. For convenience in checking the result we have generated data according to $y = 1 + 2x + 3x^2 + \cdots$. This series is truncated depending on the choice of NTERM, and some Gaussian noise is added to simulate realistic data. The sig[i] are taken as constant errors. lfit is called three times to fit the same data. The first time lista[i] is set to i, so that the fitted parameters should be returned in the order a[1] $\approx$ 1.0, a[2] $\approx$ 2.0, a[3] $\approx$ 3.0. Then, as a test of the lista feature, which determines which parameters are to be fit and in which order, the array lista[i] is reversed. Finally, the fit is restricted to odd-numbered parameters, while even-numbered parameters are fixed. In this case the elements of the covariance matrix associated with fixed parameters should be zero. In xlfit.c, we have set NTERM=3 to fit a quadratic. You may wish to try something larger.

```
/* Driver for routine LFIT */

#include <stdio.h>
#include <math.h>
#include "nr.h"
#include "nrutil.h"

#define NPT 100
#define SPREAD 0.1
#define NTERM 3

void funcs(x,afunc,mma)
float x,*afunc;
int mma;
{
    int i;

    afunc[1]=1.0;
    for (i=2;i<=mma;i++) afunc[i]=x*afunc[i-1];
}

main()
{
    int i,ii,idum=(-911),j,mfit,*lista;
    float chisq,*a,*x,*y,*sig,**covar;

    lista=ivector(1,NTERM);
    a=vector(1,NTERM);
    x=vector(1,NPT);
    y=vector(1,NPT);
    sig=vector(1,NPT);
    covar=matrix(1,NTERM,1,NTERM);
    for (i=1;i<=NPT;i++) {
        x[i]=0.1*i;
        y[i]=NTERM;
        for (j=NTERM-1;j>=1;j--)
            y[i]=j+y[i]*x[i];
        y[i] += SPREAD*gasdev(&idum);
        sig[i]=SPREAD;
    }
    mfit=NTERM;
    for (i=1;i<=mfit;i++) lista[i]=i;
    lfit(x,y,sig,NPT,a,NTERM,lista,mfit,covar,&chisq,funcs);
    printf("\n%11s %21s\n","parameter","uncertainty");
    for (i=1;i<=NTERM;i++)
        printf("  a[%1d] = %8.6f %12.6f\n",
            i,a[i],sqrt(covar[i][i]));
    printf("chi-squared = %12f\n",chisq);
    printf("full covariance matrix\n");
    for (i=1;i<=NTERM;i++) {
        for (j=1;j<=NTERM;j++) printf("%12f",covar[i][j]);
        printf("\n");
    }
    printf("\npress RETURN to continue...\n");
    getchar();
    /* Now test the LISTA feature */
    for (i=1;i<=NTERM;i++) lista[i]=NTERM+1-i;
```

```
lfit(x,y,sig,NPT,a,NTERM,lista,mfit,covar,&chisq,funcs);
printf("\n%11s %21s\n","parameter","uncertainty");
for (i=1;i<=NTERM;i++)
    printf("   a[%1d] = %8.6f %12.6f\n",
        i,a[i],sqrt(covar[i][i]));
printf("chi-squared = %12f\n",chisq);
printf("full covariance matrix\n");
for (i=1;i<=NTERM;i++) {
    for (j=1;j<=NTERM;j++) printf("%12f",covar[i][j]);
    printf("\n");
}
printf("\npress RETURN to continue...\n");
getchar();
/* Now check results of restricting fit parameters */
ii=1;
for (i=1;i<=NTERM;i++)
    if ((i % 2) == 1) lista[ii++]=i;
mfit=ii-1;
lfit(x,y,sig,NPT,a,NTERM,lista,mfit,covar,&chisq,funcs);
printf("\n%11s %21s\n","parameter","uncertainty");
for (i=1;i<=NTERM;i++)
    printf("   a[%1d] = %8.6f %12.6f\n",
        i,a[i],sqrt(covar[i][i]));
printf("chi-squared = %12f\n",chisq);
printf("full covariance matrix\n");
for (i=1;i<=NTERM;i++) {
    for (j=1;j<=NTERM;j++) printf("%12f",covar[i][j]);
    printf("\n");
}
printf("\n");
free_matrix(covar,1,NTERM,1,NTERM);
free_vector(sig,1,NPT);
free_vector(y,1,NPT);
free_vector(x,1,NPT);
free_vector(a,1,NTERM);
free_ivector(lista,1,NTERM);
}
```

`covsrt` is used in conjunction with `lfit` (and later with the routine `svdfit`) to redistribute the covariance matrix `covar` so that it represents the true order of coefficients, rather than the order in which they were fit. In sample routine `xcovsrt.c` an artificial 10×10 covariance matrix `covar[i][j]` is created, which is all zeros except for the upper left 5×5 section, for which the elements are `covar[i][j]`=i+j-1. Then three tests are performed.

1. By setting `lista[i]` $= 2i$ for $i = 1,\ldots,5$ and `mfit`=5, we spread the elements so that alternate elements are zero.

2. By taking `lista[i]` $=$ `mfit` $+ 1 - i$ for $i = 1,\ldots,5$ we put the elements in reverse order, but leave them in an upper left-hand block.

3. With `lista[i]` $= 12 - 2i$ for $i = 1,\ldots,5$ we both spread and reverse the elements.

```
/* Driver for routine COVSRT */

#include <stdio.h>
```

```
#include "nr.h"
#include "nrutil.h"

#define MA 10
#define MFIT 5

main()
{
    int i,j,*lista;
    float **covar;

    lista=ivector(1,MFIT);
    covar=matrix(1,MA,1,MA);
    for (i=1;i<=MA;i++)
        for (j=1;j<=MA;j++) {
            covar[i][j]=0.0;
            if ((i <= 5) && (j <= 5))
                covar[i][j]=i+j-1;
        }
    printf("\noriginal matrix\n");
    for (i=1;i<=MA;i++) {
        for (j=1;j<=MA;j++) printf("%4.1f",covar[i][j]);
        printf("\n");
    }
    printf("press RETURN to continue...\n");
    getchar();
    /* Test 1 - spread by 2 */
    printf("\nTest #1 - spread by two\n");
    for (i=1;i<=MFIT;i++) lista[i]=2*i;
    covsrt(covar,MA,lista,MFIT);
    for (i=1;i<=MA;i++) {
        for (j=1;j<=MA;j++) printf("%4.1f",covar[i][j]);
        printf("\n");
    }
    printf("press RETURN to continue...\n");
    getchar();
    /* Test 2 - reverse */
    printf("\nTest #2 - reverse\n");
    for (i=1;i<=MA;i++)
        for (j=1;j<=MA;j++) {
            covar[i][j]=0.0;
            if ((i <= 5) && (j <= 5)) covar[i][j]=i+j-1;
        }
    for (i=1;i<=MFIT;i++) lista[i]=MFIT+1-i;
    covsrt(covar,MA,lista,MFIT);
    for (i=1;i<=MA;i++) {
        for (j=1;j<=MA;j++) printf("%4.1f",covar[i][j]);
        printf("\n");
    }
    printf("press RETURN to continue...\n");
    getchar();
    /* Test 3 - spread and reverse */
    printf("\nTest #3 - spread and reverse\n");
    for (i=1;i<=MA;i++)
        for (j=1;j<=MA;j++) {
            covar[i][j]=0.0;
            if ((i <= 5) && (j <= 5)) covar[i][j]=i+j-1;
```

```
    }
    for (i=1;i<=MFIT;i++) lista[i]=MA+2-2*i;
    covsrt(covar,MA,lista,MFIT);
    for (i=1;i<=MA;i++) {
        for (j=1;j<=MA;j++) printf("%4.1f",covar[i][j]);
        printf("\n");
    }
    free_matrix(covar,1,MA,1,MA);
    free_ivector(lista,1,MFIT);
}
```

Routine svdfit is recommended in preference to lfit for performing linear least-squares fits. The sample program xsvdfit.c puts svdfit to work on the data generated according to

$$F(x) = 1 + 2x + 3x^2 + 4x^3 + 5x^4 + \text{Gaussian noise.}$$

This data is fit first to a five-term polynomial sum, and then to a five-term Legendre polynomial sum. In each case sig[i], the measurement fluctuation in y, is taken to be constant. For the polynomial fit, the resulting coefficients should clearly have the values a[i] $\approx$ i. For Legendre polynomials the expected results are:

$$a[1] \approx 3.0$$
$$a[2] \approx 4.4$$
$$a[3] \approx 4.9$$
$$a[4] \approx 1.6$$
$$a[5] \approx 1.1$$

```
/* Driver for routine SVDFIT */

#include <stdio.h>
#include <math.h>
#include "nr.h"
#include "nrutil.h"

#define NPT 100
#define SPREAD 0.02
#define NPOL 5

main()
{
    int i,idum=(-911);
    float chisq,*x,*y,*sig,*a,*w,**cvm,**u,**v;

    x=vector(1,NPT);
    y=vector(1,NPT);
    sig=vector(1,NPT);
    a=vector(1,NPOL);
    w=vector(1,NPOL);
    cvm=matrix(1,NPOL,1,NPOL);
    u=matrix(1,NPT,1,NPOL);
    v=matrix(1,NPOL,1,NPOL);
    for (i=1;i<=NPT;i++) {
```

```
        x[i]=0.02*i;
        y[i]=1.0+x[i]*(2.0+x[i]*(3.0+x[i]*(4.0+x[i]*5.0)));
        y[i] *= (1.0+SPREAD*gasdev(&idum));
        sig[i]=y[i]*SPREAD;
    }
    svdfit(x,y,sig,NPT,a,NPOL,u,v,w,&chisq,fpoly);
    svdvar(v,NPOL,w,cvm);
    printf("\npolynomial fit:\n\n");
    for (i=1;i<=NPOL;i++)
        printf("%12.6f %s %10.6f\n",a[i]," +-",sqrt(cvm[i][i]));
    printf("\nChi-squared %12.6f\n",chisq);
    svdfit(x,y,sig,NPT,a,NPOL,u,v,w,&chisq,fleg);
    svdvar(v,NPOL,w,cvm);
    printf("\nLegendre polynomial fit:\n\n");
    for (i=1;i<=NPOL;i++)
        printf("%12.6f %s %10.6f\n",a[i]," +-",sqrt(cvm[i][i]));
    printf("\nChi-squared %12.6f\n",chisq);
    free_matrix(v,1,NPOL,1,NPOL);
    free_matrix(u,1,NPT,1,NPOL);
    free_matrix(cvm,1,NPOL,1,NPOL);
    free_vector(w,1,NPOL);
    free_vector(a,1,NPOL);
    free_vector(sig,1,NPT);
    free_vector(y,1,NPT);
    free_vector(x,1,NPT);
}
```

svdvar is used with svdfit to evaluate the covariance matrix cvm of a fit with MA parameters. In program xsvdvar.c, we provide input vector w and array v for this routine, and then calculate the covariance matrix cvm determined from them. We have also done the calculation by hand and recorded the correct results in array tru for comparison.

```
/* Driver for routine SVDVAR */

#include <stdio.h>
#include "nr.h"
#include "nrutil.h"

#define NP 6
#define MA 3

main()
{
    int i,j;
    float **cvm,**v;
    static float vtemp[NP][NP]=
        {1.0,1.0,1.0,1.0,1.0,1.0,
         2.0,2.0,2.0,2.0,2.0,2.0,
         3.0,3.0,3.0,3.0,3.0,3.0,
         4.0,4.0,4.0,4.0,4.0,4.0,
         5.0,5.0,5.0,5.0,5.0,5.0,
         6.0,6.0,6.0,6.0,6.0,6.0};
    static float w[NP+1]=
        {0.0,0.0,1.0,2.0,3.0,4.0,5.0};
    static float tru[MA][MA]=
        {1.25,2.5,3.75,
```

```
        2.5,5.0,7.5,
        3.75,7.5,11.25};

    cvm=matrix(1,MA,1,MA);
    v=convert_matrix(&vtemp[0][0],1,NP,1,NP);
    printf("\nmatrix v\n");
    for (i=1;i<=NP;i++) {
        for (j=1;j<=NP;j++) printf("%12.6f",v[i][j]);
        printf("\n");
    }
    printf("\nvector w\n");
    for (i=1;i<=NP;i++) printf("%12.6f",w[i]);
    printf("\n");
    svdvar(v,MA,w,cvm);
    printf("\ncovariance matrix from svdvar\n");
    for (i=1;i<=MA;i++) {
        for (j=1;j<=MA;j++) printf("%12.6f",cvm[i][j]);
        printf("\n");
    }
    printf("\nexpected covariance matrix\n");
    for (i=1;i<=MA;i++) {
        for (j=1;j<=MA;j++) printf("%12.6f",tru[i-1][j-1]);
        printf("\n");
    }
    free_convert_matrix(v,1,NP,1,NP);
    free_matrix(cvm,1,MA,1,MA);
}
```

Routines fpoly and fleg are used with sample program xsvdfit.c to generate the powers of x and the Legendre polynomials, respectively. In the case of fpoly, sample program xfpoly.c is used to list the powers of x generated by fpoly so that they may be checked "by eye". For fleg, the generated polynomials in program xfleg.c are compared to values from routine plgndr.

```
/* Driver for routine FPOLY */

#include <stdio.h>
#include "nr.h"
#include "nrutil.h"

#define NVAL 15
#define DX 0.1
#define NPOLY 5

main()
{
    int i,j;
    float x,*afunc;

    afunc=vector(1,NPOLY);
    printf("\n%38s\n","powers of x");
    printf("%8s %10s %9s %9s %9s %9s\n",
        "x","x**0","x**1","x**2","x**3","x**4");
    for (i=1;i<=NVAL;i++) {
        x=i*DX;
        fpoly(x,afunc,NPOLY);
        printf("%10.4f",x);
```

```
            for (j=1;j<=NPOLY;j++) printf("%10.4f",afunc[j]);
            printf("\n");
        }
        free_vector(afunc,1,NPOLY);
}

/* Driver for routine FLEG */

#include <stdio.h>
#include "nr.h"
#include "nrutil.h"

#define NVAL 5
#define DX 0.2
#define NPOLY 5

main()
{
    int i,j;
    float x,*afunc;

    afunc=vector(1,NPOLY);
    printf("\n%3s\n","Legendre polynomials");
    printf("%9s %9s %9s %9s %9s\n","n=1","n=2","n=3","n=4","n=5");
    for (i=1;i<=NVAL;i++) {
        x=i*DX;
        fleg(x,afunc,NPOLY);
        printf("x =%5.2f\n",x);
        for (j=1;j<=NPOLY;j++) printf("%10.4f",afunc[j]);
        printf("  routine FLEG\n");
        for (j=1;j<=NPOLY;j++) printf("%10.4f",plgndr(j-1,0,x));
        printf("  routine PLGNDR\n\n");
    }
    free_vector(afunc,1,NPOLY);
}
```

mrqmin is used along with mrqcof to perform nonlinear least-squares fits with the Levenberg-Marquardt method. The artificial data used to try it in sample program xmrqmin.c is computed as the sum of two Gaussians plus noise:

$$y[i] = a[1]\exp\{-[(x[i] - a[2])/a[3]]^2\} \\ + a[4]\exp\{-[(x[i] - a[5])/a[6]]^2\} + \text{noise}.$$

The a[i] are assigned at the beginning of the program, as are the initial guesses gues[i] for these parameters to be used in initiating the fit. Also initialized for the fit are lista[i]=i for i=1,..,mfit to specify that all six of the parameters are to be fit. On the first call to mrqmin, alamda=-1 to initialize. Then a loop is entered in which mrqmin is iterated while testing successive values of chi-squared chisq. When chisq changes by less than 0.1 on two consecutive iterations, the fit is considered complete, and mrqmin is called one final time with alamda=0.0 so that array covar will return the covariance matrix. Uncertainties are derived from the square roots of the diagonal elements of covar. Expected results for the parameters are, of course, the values used to generate the "data" in the first place. The procedure fgauss is used to generate the y[i].

```
/* Driver for routine MRQMIN */

#include <stdio.h>
#include <math.h>
#include "nr.h"
#include "nrutil.h"

#define NPT 100
#define MA 6
#define SPREAD 0.001
#define SQR(a) ((a)*(a))

main()
{
    int i,idum=(-911),itst,j,k,mfit,*lista;
    float alamda,chisq,ochisq,*x,*y,*sig,**covar,**alpha;
    static float a[MA+1]=
        {0.0,5.0,2.0,3.0,2.0,5.0,3.0};
    static float gues[MA+1]=
        {0.0,4.5,2.2,2.8,2.5,4.9,2.8};

    lista=ivector(1,MA);
    x=vector(1,NPT);
    y=vector(1,NPT);
    sig=vector(1,NPT);
    covar=matrix(1,MA,1,MA);
    alpha=matrix(1,MA,1,MA);
    for (i=1;i<=NPT;i++) {
        x[i]=0.1*i;
        y[i]=0.0;
        for (j=1;j<=MA;j+=3)
            y[i] += a[j]*exp(-SQR((x[i]-a[j+1])/a[j+2]));
        y[i] *= (1.0+SPREAD*gasdev(&idum));
        sig[i]=SPREAD*y[i];
    }
    mfit=6;
    for (i=1;i<=mfit;i++) lista[i]=i;
    alamda = -1;
    for (i=1;i<=MA;i++) a[i]=gues[i];
    mrqmin(x,y,sig,NPT,a,MA,lista,mfit,covar,alpha,&chisq,fgauss,&alamda);
    k=1;
    itst=0;
    while (itst < 2) {
        printf("\n%s %2d %17s %10.4f %10s %9.2e\n","Iteration #",k,
            "chi-squared:",chisq,"alamda:",alamda);
        printf("%8s %8s %8s %8s %8s %8s\n",
            "a[1]","a[2]","a[3]","a[4]","a[5]","a[6]");
        for (i=1;i<=6;i++) printf("%9.4f",a[i]);
        printf("\n");
        k++;
        ochisq=chisq;
        mrqmin(x,y,sig,NPT,a,MA,lista,mfit,covar,alpha,
            &chisq,fgauss,&alamda);
        if (chisq > ochisq)
            itst=0;
        else if (fabs(ochisq-chisq) < 0.1)
            itst++;
```

```
    }
    alamda=0.0;
    mrqmin(x,y,sig,NPT,a,MA,lista,mfit,covar,alpha,&chisq,fgauss,&alamda);
    printf("\nUncertainties:\n");
    for (i=1;i<=6;i++) printf("%9.4f",sqrt(covar[i][i]));
    printf("\n");
    free_matrix(alpha,1,MA,1,MA);
    free_matrix(covar,1,MA,1,MA);
    free_vector(sig,1,NPT);
    free_vector(y,1,NPT);
    free_vector(x,1,NPT);
    free_ivector(lista,1,MA);
}
```

The nonlinear least-squares fit makes use of a vector β_k (the gradient of χ^2 in parameter-space) and α_{kl} (the Hessian of χ^2 in the same space). These quantities are produced by mrqcof, as demonstrated by sample program xmrqcof.c. The function is a sum of two Gaussians with noise added (the same function as in xmrqmin.c) and it is used twice. In the first call, lista[i]=i and mfit=6 so all six parameters are used. In the second call, lista[i]=i+3 and mfit=3 so the first three parameters are fixed and the last three, a[4]...a[6] are fit.

```
/* Driver for routine MRQCOF */

#include <stdio.h>
#include <math.h>
#include "nr.h"
#include "nrutil.h"

#define NPT 100
#define MA 6
#define SPREAD 0.1
#define SQR(a) ((a)*(a))

main()
{
    int i,j,idum=(-911),mfit,*lista;
    float chisq,*beta,*x,*y,*sig,**covar,**alpha;
    static float a[MA+1]=
        {0.0,5.0,2.0,3.0,2.0,5.0,3.0};
    static float gues[MA+1]=
        {0.0,4.9,2.1,2.9,2.1,4.9,3.1};

    lista=ivector(1,MA);
    beta=vector(1,MA);
    x=vector(1,NPT);
    y=vector(1,NPT);
    sig=vector(1,NPT);
    covar=matrix(1,MA,1,MA);
    alpha=matrix(1,MA,1,MA);
    /* First try sum of two gaussians */
    for (i=1;i<=NPT;i++) {
        x[i]=0.1*i;
        y[i]=0.0;
        y[i] += a[1]*exp(-SQR((x[i]-a[2])/a[3]));
        y[i] += a[4]*exp(-SQR((x[i]-a[5])/a[6]));
        y[i] *= (1.0+SPREAD*gasdev(&idum));
```

```
            sig[i]=SPREAD*y[i];
        }
    mfit=MA;
    for (i=1;i<=mfit;i++) lista[i]=i;
    for (i=1;i<=MA;i++) a[i]=gues[i];
    mrqcof(x,y,sig,NPT,a,MA,lista,mfit,alpha,beta,&chisq,fgauss);
    printf("\nmatrix alpha\n");
    for (i=1;i<=MA;i++) {
        for (j=1;j<=MA;j++) printf("%12.4f",alpha[i][j]);
        printf("\n");
    }
    printf("vector beta\n");
    for (i=1;i<=MA;i++) printf("%12.4f",beta[i]);
    printf("\nchi-squared: %12.4f\n\n",chisq);
    /* Next fix one line and improve the other */
    for (i=1;i<=3;i++) lista[i]=i+3;
    mfit=3;
    for (i=1;i<=MA;i++) a[i]=gues[i];
    mrqcof(x,y,sig,NPT,a,MA,lista,mfit,alpha,beta,&chisq,fgauss);
    printf("matrix alpha\n");
    for (i=1;i<=mfit;i++) {
        for (j=1;j<=mfit;j++) printf("%12.4f",alpha[i][j]);
        printf("\n");
    }
    printf("vector beta\n");
    for (i=1;i<=mfit;i++) printf("%12.4f",beta[i]);
    printf("\nchi-squared: %12.4f\n\n",chisq);
    free_matrix(alpha,1,MA,1,MA);
    free_matrix(covar,1,MA,1,MA);
    free_vector(sig,1,NPT);
    free_vector(y,1,NPT);
    free_vector(x,1,NPT);
    free_vector(beta,1,MA);
    free_ivector(lista,1,MA);
}
```

fgauss is an example of the type of procedure that must be supplied to mrqfit in order to fit a user-defined function, in this case the sum of Gaussians. fgauss calculates both the function, and its derivative with respect to each adjustable parameter in a fairly compact fashion. The sample program xfgauss.c calculates the same quantities in a more pedantic fashion, just to be sure we got everything right.

```
/* Driver for routine FGAUSS */

#include <stdio.h>
#include <math.h>
#include "nr.h"

#define NPT 3
#define NLIN 2
#define NA 3*NLIN
#define SQR(a) ((a)*(a))

main()
{
    int i,j;
    float e1,e2,f,x,y;
```

```
static float a[NA+1]={0.0,3.0,0.2,0.5,1.0,0.7,0.3};
float dyda[NA+1],df[NA+1];

printf("\n%6s %8s %8s %7s %7s %7s %7s %7s\n",
    "x","y","dyda1","dyda2","dyda3","dyda4","dyda5","dyda6");
for (i=1;i<=NPT;i++) {
    x=0.3*i;
    fgauss(x,a,&y,dyda,NA);
    e1=exp(-SQR((x-a[2])/a[3]));
    e2=exp(-SQR((x-a[5])/a[6]));
    f=a[1]*e1+a[4]*e2;
    df[1]=e1;
    df[4]=e2;
    df[2]=a[1]*e1*2.0*(x-a[2])/(a[3]*a[3]);
    df[5]=a[4]*e2*2.0*(x-a[5])/(a[6]*a[6]);
    df[3]=a[1]*e1*2.0*SQR(x-a[2])/(a[3]*a[3]*a[3]);
    df[6]=a[4]*e2*2.0*SQR(x-a[5])/(a[6]*a[6]*a[6]);
    printf("from FGAUSS\n");
    printf("%8.4f %8.4f",x,y);
    for (j=1;j<=6;j++) printf("%8.4f",dyda[j]);
    printf("\nindependent calc.\n");
    printf("%8.4f %8.4f",x,f);
    for (j=1;j<=6;j++) printf("%8.4f",df[j]);
    printf("\n\n");
    }
}
```

medfit is a procedure illustrating a more "robust" way of fitting. It performs a fit of data to a straight line, but instead of using the least-squares criterion for figuring the merit of a fit, it uses the least-absolute-deviation. For comparison, sample routine xmedfit.c fits lines to a noisy linear data set, using first the least-squares routine fit, and then the least-absolute-deviation routine medfit. You may be interested to see if you can figure out what mean value of absolute deviation you expect for data with gaussian noise of amplitude SPREAD.

```
/* Driver for routine MEDFIT */

#include <stdio.h>
#include "nr.h"
#include "nrutil.h"

#define NPT 100
#define SPREAD 0.1
#define NDATA NPT

main()
{
    int i,idum=(-1984),mwt=1;
    float a,abdev,b,chi2,q,siga,sigb;
    float *x,*y,*sig;

    x=vector(1,NDATA);
    y=vector(1,NDATA);
    sig=vector(1,NDATA);
    for (i=1;i<=NPT;i++) {
        x[i]=0.1*i;
        y[i] = -2.0*x[i]+1.0+SPREAD*gasdev(&idum);
```

```
        sig[i]=SPREAD;
    }
    fit(x,y,NPT,sig,mwt,&a,&b,&siga,&sigb,&chi2,&q);
    printf("\nAccording to routine FIT the result is:\n");
    printf("    a =  %8.4f    uncertainty:  %8.4f\n",a,siga);
    printf("    b =  %8.4f    uncertainty:  %8.4f\n",b,sigb);
    printf("    chi-squared:   %8.4f  for   %4d  points\n",chi2,NPT);
    printf("    goodness-of-fit:  %8.4f\n",q);
    printf("\nAccording to routine MEDFIT the result is:\n");
    medfit(x,y,NPT,&a,&b,&abdev);
    printf("    a =  %8.4f\n",a);
    printf("    b =  %8.4f\n",b);
    printf("    absolute deviation (per data point): %8.4f\n",abdev);
    printf("    (note: gaussian SPREAD is %8.4f)\n",SPREAD);
    free_vector(sig,1,NDATA);
    free_vector(y,1,NDATA);
    free_vector(x,1,NDATA);
}
```

`rofunc` is an auxiliary function for `medfit`. It evaluates the quantity

$$\sum_{i=1}^{N} x_i \, \text{sgn}(y_i - a - bx_i)$$

given arrays x_i and y_i. Data are communicated to and from `rofunc` primarily through global variables, but the value of the sum above is returned as the value of `rofunc(b)`. `abdev` is the summed absolute deviation, and `aa` (listed below as A) is given the value which minimizes `abdev`. Our results for these quantities are:

B	A	ROFUNC	ABDEV
-2.10	1.51	245.40	25.38
-2.08	1.41	242.20	20.54
-2.06	1.30	242.20	15.69
-2.04	1.20	234.20	10.94
-2.02	1.09	193.20	6.61
-2.00	1.00	22.00	4.07
-1.98	.89	-199.20	5.78
-1.96	.79	-237.40	10.37
-1.94	.68	-246.40	15.25
-1.92	.58	-246.40	20.18
-1.90	.48	-248.40	25.13

```
/* Driver for routine ROFUNC */

#include <stdio.h>
#include "nr.h"
#include "nrutil.h"

#define SPREAD 0.05
#define NDATA 100

int ndatat=0;      /* defining declaration */
float *xt=0,*yt=0,aa=0.0,abdevt=0.0;      /* defining declaration */

main()
{
    int i,idum=(-11);
```

```
float b,rf,*x,*y;

x=vector(1,NDATA);
y=vector(1,NDATA);
ndatat=NDATA;
xt=x;
yt=y;
for (i=1;i<=NDATA;i++) {
    x[i]=0.1*i;
    y[i] = -2.0*x[i]+1.0+SPREAD*gasdev(&idum);
}
printf("%9s %9s %12s %10s\n","b","a","ROFUNC","ABDEVT");
for (i = -5;i<=5;i++) {
    b = -2.0+0.02*i;
    rf=rofunc(b);
    printf("%10.2f %9.2f %11.2f %10.2f\n",
        b,aa,rf,abdevt);
}
free_vector(y,1,NDATA);
free_vector(x,1,NDATA);
}
```

Chapter 15: Ordinary Differential Equations

*Chapter 15 of Numerical Recipes deals with the integration of ordinary differential equations, restricting its attention specifically to initial-value problems. Three practical methods are introduced: 1) Runge-Kutta methods (*rk4, rkdumb, rkqc,* and* odeint*), 2) Richardson extrapolation and the Bulirsch-Stoer method (*bsstep, mmid, rzextr, pzextr*), 3) predictor-corrector methods. In general, for applications not demanding high precision, and where convenience is paramount, the fourth-order Runge-Kutta with adaptive step-size control is recommended. For higher precision applications, the Bulirsch-Stoer method dominates. The predictor-corrector methods are covered because of their history of widespread use, but are not regarded (by us) as having an important role today. (For a possible exception to this strong statement, see* Numerical Recipes.*)*

$$\star \quad \star \quad \star \quad \star$$

Routine rk4 advances the solution vector y[n] of a set of ordinary differential equations over a single small interval h in x using the fourth-order Runge-Kutta method. The operation is shown by sample program xrk4.c for an array of four variables y[1],...,y[4]. The first-order differential equations satisfied by these variables are specified by the accompanying routine derivs, and are simply the equations describing the first four Bessel functions $J_0(x), \ldots, J_3(x)$. The y's are initialized to the values of these functions at $x = 1.0$. Note that the values of dydx are also initialized at $x = 1.0$, because rk4 uses the values of dydx before its first call to derivs. The reason for this is discussed in the text. The sample program calls rk4 with h (the step-size) set to various values from 0.2 to 1.0, so that you can see how well rk4 can do even with quite sizeable steps.

```
/* Driver for routine RK4 */

#include <stdio.h>
#include "nr.h"
#include "nrutil.h"

#define N 4

void derivs(x,y,dydx)
float x,y[],dydx[];
{
    dydx[1] = -y[2];
    dydx[2]=y[1]-(1.0/x)*y[2];
    dydx[3]=y[2]-(2.0/x)*y[3];
    dydx[4]=y[3]-(3.0/x)*y[4];
}
```

```
main()
{
    int i,j;
    float h,x=1.0,*y,*dydx,*yout;

    y=vector(1,N);
    dydx=vector(1,N);
    yout=vector(1,N);
    y[1]=bessj0(x);
    y[2]=bessj1(x);
    y[3]=bessj(2,x);
    y[4]=bessj(3,x);
    dydx[1] = -y[2];
    dydx[2]=y[1]-y[2];
    dydx[3]=y[2]-2.0*y[3];
    dydx[4]=y[3]-3.0*y[4];
    printf("\n%16s %5s %12s %12s %12s\n",
        "Bessel function:","j0","j1","j3","j4");
    for (i=1;i<=5;i++) {
        h=0.2*i;
        rk4(y,dydx,N,x,h,yout,derivs);
        printf("\nfor a step size of: %6.2f\n",h);
        printf("%12s","rk4:");
        for (j=1;j<=4;j++) printf(" %12.6f",yout[j]);
        printf("\n%12s %12.6f %12.6f %12.6f %12.6f\n","actual:",
            bessj0(x+h),bessj1(x+h),bessj(2,x+h),bessj(3,x+h));
    }
    free_vector(yout,1,N);
    free_vector(dydx,1,N);
    free_vector(y,1,N);
}
```

rkdumb is an extension of rk4 which allows you to integrate over larger intervals. It is "dumb" in the sense that it has no adaptive step-size determination, and no code to estimate errors. Sample program xrkdumb.c works with the same functions and derivatives as the previous program, but integrates from $x1=1.0$ to $x2=20.0$, breaking the interval into NSTEP=150 equal steps. The variables vstart[1],...,vstart[4] which become the starting values of the y's, are initialized as before, but their derivatives this time are not initialized; rkdumb takes care of that. This time only the results for the fourth variable $J_3(x)$ are listed, and only every tenth value is given.

```
/* Driver for routine RKDUMB */

#include <stdio.h>
#include "nr.h"
#include "nrutil.h"

#define NVAR 4
#define NSTEP 150

void derivs(x,y,dydx)
float x,y[],dydx[];
{
    dydx[1] = -y[2];
    dydx[2]=y[1]-(1.0/x)*y[2];
    dydx[3]=y[2]-(2.0/x)*y[3];
```

```
    dydx[4]=y[3]-(3.0/x)*y[4];
}

extern float **y,*xx;     /*referencing declaration */

main()
{
    int j;
    float x1,x2,*vstart;

    vstart=vector(1,NVAR);
    /* Note: The arrays xx and y must have indices up to NSTEP+1 */
    xx=vector(1,NSTEP+1);
    y=matrix(1,NVAR,1,NSTEP+1);
    x1=1.0;
    x2=20.0;
    vstart[1]=bessj0(x1);
    vstart[2]=bessj1(x1);
    vstart[3]=bessj(2,x1);
    vstart[4]=bessj(3,x1);
    rkdumb(vstart,NVAR,x1,x2,NSTEP,derivs);
    printf("%8s %17s %10s\n","x","integrated","bessj3");
    for (j=10;j<=NSTEP;j+=10)
        printf("%10.4f %14.6f %12.6f\n",
            xx[j],y[4][j],bessj(3,xx[j])));
    free_matrix(y,1,NVAR,1,NSTEP+1);
    free_vector(xx,1,NSTEP+1);
    free_vector(vstart,1,NVAR);
}
```

rkqc performs a single step of fifth-order Runge-Kutta integration, this time with monitoring of local truncation error and corresponding step-size adjustment. Its sample program xrkqc.c is similar to that for routine rk4, using four Bessel functions as the example, and starting the integration at $x = 1.0$. However, on each pass a value is set for eps, the desired accuracy, and the trial value htry for the interval size is set to 0.1. For the first few passes, eps is not too demanding and htry may be perfectly adequate. As eps becomes smaller, the routine will be forced to diminish h and return smaller values of hdid and hnext. Our results (in single precision) are:

eps	htry	hdid	hnext
.3679E+00	.10	.100000	.400000
.1353E+00	.10	.100000	.354954
.4979E-01	.10	.100000	.287823
.1832E-01	.10	.100000	.233879
.6738E-02	.10	.100000	.190423
.2479E-02	.10	.100000	.155293
.9119E-03	.10	.100000	.126883
.3355E-03	.10	.100000	.103845
.1234E-03	.10	.073460	.066323
.4540E-04	.10	.034162	.031216
.1670E-04	.10	.028686	.025915
.6144E-05	.10	.011732	.010733
.2260E-05	.10	.010758	.009784
.8315E-06	.10	.004284	.003911
.3059E-06	.10	.004460	.004035

```
/* Driver for routine RKQC */

#include <stdio.h>
#include <math.h>
#include "nr.h"
#include "nrutil.h"

#define N 4

void derivs(x,y,dydx)
float x,y[],dydx[];
{
    dydx[1] = -y[2];
    dydx[2]=y[1]-(1.0/x)*y[2];
    dydx[3]=y[2]-(2.0/x)*y[3];
    dydx[4]=y[3]-(3.0/x)*y[4];
}

main()
{
    int i;
    float eps,hdid,hnext,htry,x=1.0,*y,*dydx,*yscal;

    y=vector(1,N);
    dydx=vector(1,N);
    yscal=vector(1,N);
    y[1]=bessj0(x);
    y[2]=bessj1(x);
    y[3]=bessj(2,x);
    y[4]=bessj(3,x);
    dydx[1] = -y[2];
    dydx[2]=y[1]-y[2];
    dydx[3]=y[2]-2.0*y[3];
    dydx[4]=y[3]-3.0*y[4];
    for (i=1;i<=N;i++) yscal[i]=1.0;
    htry=0.1;
    printf("%10s %11s %12s %13s\n","eps","htry","hdid","hnext");
    for (i=1;i<=15;i++) {
        eps=exp((double) -i);
        rkqc(y,dydx,N,&x,htry,eps,yscal,&hdid,&hnext,derivs);
        printf("%13f %8.2f %14.6f %12.6f \n",eps,htry,hdid,hnext);
    }
    free_vector(yscal,1,N);
    free_vector(dydx,1,N);
    free_vector(y,1,N);
}
```

The full driver routine for rkqc, which provides Runge-Kutta integration over large intervals with adaptive step-size control, is odeint. It plays the same role for rkqc that rkdumb plays for rk4. Integration is performed on four Bessel functions from x1=1.0 to x2=10.0, with an accuracy eps=1.0e-4. Independent of the values of step-size actually used by odeint, intermediate values will be recorded only at intervals greater than dxsav. The sample program returns values of $J_3(x)$ for checking against actual values produced by bessj. It also records how many steps were successful, and how many were "bad". Bad steps are redone, and indicate no extra loss in accuracy. At the same time, they do represent a loss in efficiency, so that an excessive number of bad steps should initiate an investigation.

```
/* Driver for ODEINT */

#include <stdio.h>
#include "nr.h"
#include "nrutil.h"

#define N 4

extern float dxsav,*xp,**yp;     /* defined in ODEINT.C */
extern int kount,kmax;          /* defined in ODEINT.C */

void derivs(x,y,dydx)
float x,y[],dydx[];
{
    dydx[1] = -y[2];
    dydx[2]=y[1]-(1.0/x)*y[2];
    dydx[3]=y[2]-(2.0/x)*y[3];
    dydx[4]=y[3]-(3.0/x)*y[4];
}

main()
{
    int i,nbad,nok;
    float eps,h1,hmin,x1,x2,*ystart;

    ystart=vector(1,N);
    xp=vector(1,200);
    yp=matrix(1,10,1,200);
    x1=1.0;
    x2=10.0;
    ystart[1]=bessj0(x1);
    ystart[2]=bessj1(x1);
    ystart[3]=bessj(2,x1);
    ystart[4]=bessj(3,x1);
    eps=1.0e-4;
    h1=0.1;
    hmin=0.0;
    kmax=100;
    dxsav=(x2-x1)/20.0;
    odeint(ystart,N,x1,x2,eps,h1,hmin,&nok,&nbad,derivs,rkqc);
    printf("\n%s %13s %3d\n","successful steps:"," ",nok);
    printf("%s %20s %3d\n","bad steps:"," ",nbad);
    printf("\n%s %3d\n","stored intermediate values:    ",kount);
    printf("\n%8s %18s %15s\n","x","integral","bessj(3,x)");
    for (i=1;i<=kount;i++)
        printf("%10.4f %16.6f %14.6f\n",
            xp[i],yp[4][i],bessj(3,xp[i]));
    free_matrix(yp,1,10,1,200);
    free_vector(xp,1,200);
    free_vector(ystart,1,N);
}
```

The modified midpoint routine `mmid` is presented in *Numerical Recipes* primarily as a component of the more powerful Bulirsch-Stoer routine. It integrates variables over an interval `htot` through a sequence of much smaller steps. Sample routine `xmmid.c` takes the number of subintervals i to be $5, 10, 15, \ldots, 50$ so that we can witness any improvements in accuracy that may occur. The values of the four Bessel functions are compared with the results of the integrations.

```c
/* Driver for routine MMID */

#include <stdio.h>
#include "nr.h"
#include "nrutil.h"

#define NVAR 4
#define X1 1.0
#define HTOT 0.5

void derivs(x,y,dydx)
float x,y[],dydx[];
{
    dydx[1] = -y[2];
    dydx[2]=y[1]-(1.0/x)*y[2];
    dydx[3]=y[2]-(2.0/x)*y[3];
    dydx[4]=y[3]-(3.0/x)*y[4];
}

main()
{
    int i;
    float b1,b2,b3,b4,xf,*y,*yout,*dydx;

    y=vector(1,NVAR);
    yout=vector(1,NVAR);
    dydx=vector(1,NVAR);
    y[1]=bessj0(X1);
    y[2]=bessj1(X1);
    y[3]=bessj(2,X1);
    y[4]=bessj(3,X1);
    dydx[1] = -y[2];
    dydx[2]=y[1]-y[2];
    dydx[3]=y[2]-2.0*y[3];
    dydx[4]=y[3]-3.0*y[4];
    xf=X1+HTOT;
    b1=bessj0(xf);
    b2=bessj1(xf);
    b3=bessj(2,xf);
    b4=bessj(3,xf);
    printf("First four Bessel functions:\n");
    for (i=5;i<=50;i += 5) {
        mmid(y,dydx,NVAR,X1,HTOT,i,yout,derivs);
        printf("\n%s %5.2f %s %5.2f %s %2d %s \n",
            "x=",X1," to ",X1+HTOT," in ",i," steps");
        printf("%14s %9s\n","integration","bessj");
        printf("%12.6f %12.6f\n",yout[1],b1);
        printf("%12.6f %12.6f\n",yout[2],b2);
        printf("%12.6f %12.6f\n",yout[3],b3);
        printf("%12.6f %12.6f\n",yout[4],b4);
        printf("\nPress RETURN to continue...\n");
        getchar();
    }
```

```
    free_vector(dydx,1,NVAR);
    free_vector(yout,1,NVAR);
    free_vector(y,1,NVAR);
}
```

The Bulirsch-Stoer method, illustrated by routine bsstep, is the integrator of choice for higher accuracy calculations on smooth functions. An interval h is broken into finer and finer steps, and the results of integration are extrapolated to zero step-size. The extrapolation is via a rational function with rzextr. bsstep monitors local truncation error and adjusts the step-size appropriately, to keep errors below eps. From an external point of view, bsstep operates exactly as does rkqc: it has the same arguments and in the same order. Consequently it can be used in place of rkqc in routine odeint, allowing more efficient integration over large regions of x. For this reason, the sample program xbsstep.c is of the same form used to demonstrate rkqc.

```
/* Driver for routine BSSTEP */

#include <stdio.h>
#include <math.h>
#include "nr.h"
#include "nrutil.h"

#define N 4
#define IMAX 11
#define NUSE 7

void derivs(x,y,dydx)
float x,y[],dydx[];
{
    dydx[1] = -y[2];
    dydx[2]=y[1]-(1.0/x)*y[2];
    dydx[3]=y[2]-(2.0/x)*y[3];
    dydx[4]=y[3]-(3.0/x)*y[4];
}

main()
{
    int i;
    float eps,hdid,hnext,htry,x=1.0,*y,*dydx,*yscal;

    y=vector(1,N);
    dydx=vector(1,N);
    yscal=vector(1,N);
    y[1]=bessj0(x);
    y[2]=bessj1(x);
    y[3]=bessj(2,x);
    y[4]=bessj(3,x);
    dydx[1] = -y[2];
    dydx[2]=y[1]-y[2];
    dydx[3]=y[2]-2.0*y[3];
    dydx[4]=y[3]-3.0*y[4];
    for (i=1;i<=N;i++) yscal[i]=1.0;
    htry=1.0;
    printf("\n%10s %10s %12s %12s \n","eps","htry","hdid","hnext");
    for (i=1;i<=15;i++) {
        eps=exp((double) -i);
```

```
        bsstep(y,dydx,N,&x,htry,eps,yscal,&hdid,&hnext,derivs);
        printf(" %11f %8.2f %14.6f %12.6f\n",eps,htry,hdid,hnext);
    }
    free_vector(yscal,1,N);
    free_vector(dydx,1,N);
    free_vector(y,1,N);
}
```

rzextr performs a diagonal rational function extrapolation for bsstep. It takes a sequence of interval lengths and corresponding integrated values, and extrapolates to the value the integral would have if the interval length were zero. Sample routine xrzextr.c works with a known function

$$F_n = \frac{1 - x + x^3}{(x+1)^n} \qquad n = 1, .., 4$$

We extrapolate the vector yest $= (F_1, F_2, F_3, F_4)$ given a sequence of ten values (only the last NUSE=5 of which are used). The ten values are labelled iest=1,...,10 and are evaluated at xest=1.0/iest. A call to rzextr produces extrapolated values yz, and estimated errors dy, and compares to the true values $(1.0, 1.0, 1.0, 1.0)$ at xest=0.0.

```
/* Driver for routine RZEXTR */

#include <stdio.h>
#include "nr.h"
#include "nrutil.h"

#define NV 4
#define NUSE  5
#define IMAX 10

float **d=0,*x=0;      /* defining declaration */

main()
{
    int i,iest,j;
    float dum,xest,*dy,*yest,*yz;

    dy=vector(1,NV);
    yest=vector(1,NV);
    yz=vector(1,NV);
    x=vector(1,IMAX);
    d=matrix(1,NV,1,NUSE);
    /* Feed values from a rational function */
    /* fn(x)=(1-x+x**3)/(x+1)**n */
    for (i=1;i<=IMAX;i++) {
        iest=i;
        xest=1.0/i;
        dum=1.0-xest+xest*xest*xest;
        for (j=1;j<=NV;j++) {
            dum /= (xest+1.0);
            yest[j]=dum;
        }
        rzextr(iest,xest,yest,yz,dy,NV,NUSE);
        printf("\n%s %2d %s %8.4f\n",
            "iest = ",i," xest =",xest);
        printf("Extrap. function: ");
```

```
        for (j=1;j<=NV;j++) printf("%12.6f",yz[j]);
        printf("\nEstimated error:   ");
        for (j=1;j<=NV;j++) printf("%12.6f",dy[j]);
        printf("\n");
    }
    printf("\nActual values: %15.6f %11.6f %11.6f %11.6f \n",
        1.0,1.0,1.0,1.0);
    free_matrix(d,1,NV,1,NUSE);
    free_vector(x,1,IMAX);
    free_vector(yz,1,NV);
    free_vector(yest,1,NV);
    free_vector(dy,1,NV);
}
```

pzextr is a less powerful standby for rzextr, to be used primarily when some problem crops up with the extrapolation. It performs a polynomial, rather than a rational function, extrapolation. The sample program xpzextr.c is identical to that for rzextr.

```
/* Driver for routine PZEXTR */

#include <stdio.h>
#include "nr.h"
#include "nrutil.h"

#define NV 4
#define NUSE 5
#define IMAX 10

float **d=0,*x=0;      /* defining declaration */

main()
{
    int i,iest,j;
    float dum,xest,*dy,*yest,*yz;

    dy=vector(1,NV);
    yest=vector(1,NV);
    yz=vector(1,NV);
    x=vector(1,IMAX);
    d=matrix(1,NV,1,NUSE);
    /* Feed values from a rational function */
    /* fn(x)=(1-x+x**3)/(x+1)**n */
    for (i=1;i<=IMAX;i++) {
        iest=i;
        xest=1.0/i;
        dum=1.0-xest+xest*xest*xest;
        for (j=1;j<=NV;j++) {
            dum /= (xest+1.0);
            yest[j]=dum;
        }
        pzextr(iest,xest,yest,yz,dy,NV,NUSE);
        printf("\ni = %2d",i);
        printf("\nExtrap. function:");
        for (j=1;j<=NV;j++) printf("%12.6f",yz[j]);
        printf("\nEstimated error: ");
        for (j=1;j<=NV;j++) printf("%12.6f",dy[j]);
```

```
        printf("\n");
    }
    printf("\nactual values: %14.6f %11.6f %11.6f %11.6f\n",
        1.0,1.0,1.0,1.0);
    free_matrix(d,1,NV,1,NUSE);
    free_vector(x,1,IMAX);
    free_vector(yz,1,NV);
    free_vector(yest,1,NV);
    free_vector(dy,1,NV);
}
```

Chapter 16: Two-Point Boundary Value Problems

Two-point boundary value problems, and their iterative solution, is the substance of Chapter 16 of *Numerical Recipes*. The first step is to cast the problem as a set of N coupled first-order ordinary differential equations, satisfying n_1 conditions at one boundary point, and $n_2 = N - n_1$ conditions at the other boundary point. We apply two general methods to the solutions. First are the shooting methods, typified by procedures shoot and shootf, which enforce the n_1 conditions at one boundary and set n_2 conditions freely. Then they integrate across the interval to find discrepancies with the n_2 conditions at the other end. The Newton-Raphson method is used to reduce these discrepancies by adjusting the variable parameters.

The other approach is the relaxation method in which the differential equations are replaced by finite difference equations on a grid that covers the range of interest. Routine solvde demonstrates this method, and is demonstrated "in action" by program sfroid, which uses it to compute eigenvalues of spheroidal harmonics. Since the program sfroid in *Numerical Recipes* is already self-contained, we need concern ourselves here only with shooting routines. For the purpose of comparison, we apply these routines to the same problem attacked with sfroid.

$$\star \quad \star \quad \star \quad \star$$

Procedure shoot works as described above. Demonstration program xshoot.c uses it to find eigenvalues of both prolate and oblate spheroidal harmonics. The oblate and prolate cases are handled simultaneously, although they actually involve two independent sets of three coupled first-order differential equations, one set with c^2 positive and the other with c^2 negative. The complete set of differential equations is

$$\frac{dy_1}{dx} = y_2$$

$$\frac{dy_2}{dx} = \frac{2x(m+1)y_2 - (y_3 - c^2 x^2)y_1}{(1 - x^2)}$$

$$\frac{dy_3}{dx} = 0$$

$$\frac{dy_4}{dx} = y_5$$

$$\frac{dy_5}{dx} = \frac{2x(m+1)y_5 - (y_6 + c^2 x^2)y_4}{(1 - x^2)}$$

$$\frac{dy_6}{dx} = 0$$

These are specified in procedure `derivs` which is called, in turn, by `odeint` in `shoot`. The first three equations correspond to prolate harmonics and the second three to oblate harmonics. Comparing either set of three to equation (16.4.4) in *Numerical Recipes*, you may quickly verify that y_1 and y_4 correspond to the two spheroidal harmonic solutions, y_3 and y_6 correspond to the sought-after eigenvalues (whose derivative with respect to x is of course 0), and y_2 and y_5 are intermediate variables created to change the second-order equations to coupled first-order equations.

Two other procedures are used by `shoot`. Procedure `load` sets the values of all the variables $y_1, \ldots, y_6$ at the first boundary, and `score` calculates a discrepancy vector F (which will be zero when a successful solution has been reached) at the second boundary. Each of these procedures has some interesting aspects. In `load`, y_3 and y_6 are initialized to `v[1]` and `v[2]`, values calculated in the sample program to give rough estimates of the size of the proper result. We arrived at these estimates just by looking through some tables of values. Also notice that, for example, y_1 is set to `factr` $+ y_2$`DX`. This is the same as saying that $y_1 = $ `factr` $+ (dy_1/dx)\Delta x$. The quantity `factr` comes from equation (16.4.20) in *Numerical Recipes*, and the term with DX comes from the fact that we placed the lower boundary x1 at $-1.0 + $ DX (where DX=1.0E-4) rather than at -1.0. This is because dy_2/dx and dy_5/dx cannot be evaluated exactly at $x = -1.0$. The procedure `score` follows from equation (16.4.18) in *Numerical Recipes*. For example, if $N - M$ is odd, $y_1 = 0$ at $x = 0$, but if $N - M$ is even, then $y_2 = dy_1/dx = 0$.

That more or less explains things. Now, given M, N, and c^2, sample program `xshoot.c` sets up estimates `v[1]` and `v[2]` and iterates the routine `shoot` until changes in the v are less than some preset fraction `EPS` of their size. Some values of the eigenvalues of the spheroidal harmonics are given in section 16.4 of *Numerical Recipes* if you want to check the results.

```
/* Driver for routine SHOOT */
/* Solves for eigenvalues of spheroidal harmonics. Both
prolate and oblate case are handled simultaneously, leading
to six first-order equations. Unknown to shoot, it is
actually two independent sets of three coupled equations,
one set with c^2 positive and the other with c^2 negative. */

#include <stdio.h>
#include <math.h>
#include "nr.h"

#define NVAR 6
#define N2 2
#define DELTA 1.0e-3
#define EPS 1.0e-6
#define DX 1.0e-4

float c2,factr;
int m,n;

void load(x1,v,y)
float x1,v[],y[];
{
    y[3]=v[1];
    y[2] = -(y[3]-c2)*factr/2.0/(m+1.0);
    y[1]=factr+y[2]*DX;
```

```
    y[6]=v[2];
    y[5] = -(y[6]+c2)*factr/2.0/(m+1.0);
    y[4]=factr+y[5]*DX;
}

void score(x2,y,f)
float x2,y[],f[];
{
    if ((n-m) % 2 == 0) {
        f[1]=y[2];
        f[2]=y[5];
    } else {
        f[1]=y[1];
        f[2]=y[4];
    }
}

void derivs(x,y,dydx)
float x,y[],dydx[];
{
    dydx[1]=y[2];
    dydx[3]=0.0;
    dydx[2]=(2.0*x*(m+1.0)*y[2]-(y[3]-c2*x*x)*y[1])/(1.0-x*x);
    dydx[4]=y[5];
    dydx[6]=0.0;
    dydx[5]=(2.0*x*(m+1.0)*y[5]-(y[6]+c2*x*x)*y[4])/(1.0-x*x);
}

main()
{
    int i;
    float h1,hmin,q1,x1,x2;
    float delv[3],v[3],dv[7],f[7];

    do {
        printf("Input M,N,C-Squared:   ");
        scanf("%d %d %f",&m,&n,&c2);
    } while (n < m || m < 0);
    factr=1.0;
    if (m) {
        q1=n;
        for (i=1;i<=m;i++) {
            factr *= (-0.5*(n+i)*(q1/i));
            q1 -= 1.0;
        }
    }
    v[1]=n*(n+1)-m*(m+1)+c2/2.0;
    v[2]=n*(n+1)-m*(m+1)-c2/2.0;
    delv[1]=DELTA*v[1];
    delv[2]=delv[1];
    h1=0.1;
    hmin=0.0;
    x1 = -1.0+DX;
    x2=0.0;
    printf("\n%17s %24s \n","Prolate","Oblate");
    printf("%11s %14s %10s %14s\n",
        "Mu(m,n)","Error Est.","Mu(m,n)","Error Est.");
```

```
    do {
        shoot(NVAR,v,delv,N2,x1,x2,EPS,h1,hmin,f,dv);
        printf("%12.6f %12.6f %12.6f %12.6f\n",v[1],dv[1],v[2],dv[2]);
    } while (fabs(dv[1]) > fabs(EPS*v[1]) || dv[2] > fabs(EPS*v[2]));
}
```

Another shooting method is shooting to a fitting point. More explicitly, we set values at two boundaries, from both of which we integrate toward an intermediate point. For the spheroidal harmonics, we take the endpoints, in sample program xshootf.c, to be $-1.0 + $ DX and $1.0 - $ DX, and the intermediate point to be $x = 0.0$. For clarity, we consider only prolate spheroids. The calculation is similar to that in the previous sample program, except for these details:

1. There are only three first-order differential equations in derivs because of the restriction to prolate spheroids. (Note: the oblate case requires only that we input c^2 as a negative number.)

2. There are two load routines, load1 and load2, which set values at the two boundaries. At the first boundary y[3] is initialized to v1[1], which is initially set to our crude guess of the magnitude of the eigenvalue. y[1], the spheroidal harmonic value itself, is set to factr $+ (dy_1/dx)\Delta x$, and y[2] is also set as before. At boundary two, y[3] and y[1] are given guessed values for the eigenvalue and for $y(1 - \Delta x)$ respectively. We treat the guessed eigenvalue at boundary two as independent of that at boundary one, although they ought certainly to converge to the same value. To verify this point, we make the initial guess that the values differ by 1.0 (i.e. v2[2]=v1[1]+1.0).

Sample program xshootf.c otherwise proceeds much as xshoot.c did, however with score kept at $x = 0.0$ where the solutions must match up. The procedure score has been set to a dummy operation equating F_i to y_i so that the condition of success is that the y_i all match at $x = 0$. This is discussed more fully in *Numerical Recipes*. Check the eigenvalue results against the previous routine.

```
/* Driver for routine SHOOTF */

#include <stdio.h>
#include <math.h>
#include "nr.h"

#define NVAR 3
#define N1 2
#define N2 1
#define DELTA 1.0e-3
#define EPS 1.0e-6
#define DX 1.0e-4

float c2,factr,dx;
int m;

void load1(x1,v1,y)
float x1,v1[],y[];
{
    y[3]=v1[1];
    y[2] = -(y[3]-c2)*factr/2.0/(m+1.0);
    y[1]=factr+y[2]*dx;
}
```

```
void load2(x2,v2,y)
float x2,v2[],y[];
{
    y[3]=v2[2];
    y[1]=v2[1];
    y[2]=(y[3]-c2)*y[1]/2.0/(m+1.0);
}

void score(xf,y,f)
float xf,y[],f[];
{
    int i;

    for (i=1;i<=3;i++) f[i]=y[i];
}

void derivs(x,y,dydx)
float x,y[],dydx[];
{
    dydx[1]=y[2];
    dydx[3]=0.0;
    dydx[2]=(2.0*x*(m+1.0)*y[2]-(y[3]-c2*x*x)*y[1])/(1.0-x*x);
}

main()
{
    int i,n;
    float h1,hmin,q1,x1,x2,xf;
    float v1[2],delv1[2],dv1[2],v2[3],delv2[3],dv2[3],f[4];

    do {
        printf("Input M,N,C-SQUARED: ");
        scanf("%d %d %f",&m,&n,&c2);
    } while (n < m || m < 0);
    dx=DX;
    factr=1.0;
    if (m) {
        q1=n;
        for (i=1;i<=m;i++) {
            factr *= (-0.5*(n+i)*(q1/i));
            q1 -= 1.0;
        }
    }
    v1[1]=n*(n+1)-m*(m+1)+c2/2.0;
    v2[1]=((n-m)%2 == 0 ? factr : -factr);
    v2[2]=v1[1]+1.0;
    delv1[1]=DELTA*v1[1];
    delv2[1]=DELTA*factr;
    delv2[2]=delv1[1];
    h1=0.1;
    hmin=0.0;
    x1 = -1.0+DX;
    x2=1.0-DX;
    xf=0.0;
    printf("\n %26s %20s %19s\n","mu(-1)","y(1-dx)","mu(+1)");
    for (dv1[1]=v1[1];fabs(dv1[1]) > fabs(EPS*v1[1]);) {
```

```
        shootf(NVAR,v1,v2,delv1,delv2,N1,N2,x1,x2,
            xf,EPS,h1,hmin,f,dv1,dv2);
        printf("\n%6s %20.6f %20.6f %20.6f\n",
            "v ",v1[1],v2[1],v2[2]);
        printf("%6s %20.6f %20.6f %20.6f\n",
            "dv",dv1[1],dv2[1],dv2[2]);
    }
}
```

Chapter 17: Partial Differential Equations

Several methods for solving partial differential equations by numerical means are treated in Chapter 17 of Numerical Recipes. All are finite differencing methods, including forward time centered space differencing, the Lax method, staggered leapfrog differencing, the two-step Lax-Wendroff scheme, the Crank-Nicholson method, Fourier analysis and cyclic reduction (FACR), Jacobi's method, the Gauss-Seidel method, simultaneous over-relaxation (SOR) with and without Chebyshev acceleration, and operator splitting methods as exemplified by the alternating direction implicit (ADI) method. There are so many methods, in fact, that we have not provided each topic with a procedure of its own. In many cases the nature of such procedures follows naturally from the description. In other cases, you will have to consult other references. The procedures that do appear in the chapter, sor *and* adi, *show two of the more useful and efficient methods for elliptic equations in application.*

★ ★ ★ ★

Procedure sor incorporates simultaneous over-relaxation with Chebyshev acceleration to solve an elliptic partial differential equation. As input it accepts six arrays of coefficients, an estimate of the spectral radius of Jacobi iteration, and a trial solution which is often just set to zero over the solution grid. In program xsor.c the method is applied to the model problem

$$\frac{\partial^2 u}{\partial x^2} + \frac{\partial^2 u}{\partial y^2} = \rho$$

which is treated as the relaxation problem

$$\frac{\partial u}{\partial t} = \frac{\partial^2 u}{\partial x^2} + \frac{\partial^2 u}{\partial y^2} - \rho$$

Using FTCS differencing, this becomes

$$u^n_{j+1,l} + u^n_{j-1,l} + u^n_{j,l+1} + u^n_{j,l-1} - 4u^{n+1}_{j,l} = \rho_{jl}\Delta^2$$

(The notation is explained in Chapter 17 of *Numerical Recipes*.) This is a simple form of the general difference equation to which sor may be applied, with

$$A_{jl} = B_{jl} = C_{jl} = D_{jl} = 1.0 \text{ and } E_{jl} = -4.0$$

for all j and l. The starting guess for u is $u_{jl} = 0.0$ for all j, l. For a source function $F_{j,l}$ we took $F_{j,l} = 0.0$ except directly in the center of the grid where $F(\text{midl}, \text{midl}) = 2.0$.

218

The value of ρ_{Jacobi}, which is called $rjac$, is taken from equation (17.5.24) of *Numerical Recipes*,

$$\rho_{Jacobi} = \frac{\cos\dfrac{\pi}{J} + \left(\dfrac{\Delta x}{\Delta y}\right)^2 \cos\dfrac{\pi}{L}}{1 + \left(\dfrac{\Delta x}{\Delta y}\right)^2}$$

In this case, $j=l=$JMAX and $\Delta x = \Delta y$ so $rjac = \cos(\pi/\text{JMAX})$. A call to sor leads to the solution shown below. As a test that this is indeed a solution to the finite difference equation, the program plugs the result back into that equation, calculating

$$F_{j,l} = u_{j+1,l}^n + u_{j-1,l}^n + u_{j,l+1}^n + u_{j,l-1}^n - 4u_{j,l}^{n+1}$$

The test is whether $F_{j,l}$ is almost everywhere zero, but equal to 2.0 at the very centerpoint of the grid.

```
SOR solution grid:
 .00    .00    .00    .00    .00    .00    .00    .00    .00    .00    .00
 .00  -.02  -.04  -.06  -.08  -.09  -.08  -.06  -.04  -.02    .00
 .00  -.04  -.09  -.13  -.17  -.19  -.17  -.13  -.09  -.04    .00
 .00  -.06  -.13  -.20  -.28  -.32  -.28  -.20  -.13  -.06    .00
 .00  -.08  -.17  -.28  -.41  -.55  -.41  -.28  -.17  -.08    .00
 .00  -.09  -.19  -.32  -.55 -1.05  -.55  -.32  -.19  -.09    .00
 .00  -.08  -.17  -.28  -.41  -.55  -.41  -.28  -.17  -.08    .00
 .00  -.06  -.13  -.20  -.28  -.32  -.28  -.20  -.13  -.06    .00
 .00  -.04  -.09  -.13  -.17  -.19  -.17  -.13  -.09  -.04    .00
 .00  -.02  -.04  -.06  -.08  -.09  -.08  -.06  -.04  -.02    .00
 .00    .00    .00    .00    .00    .00    .00    .00    .00    .00    .00
```

```
/* Driver for routine SOR */

#include <stdio.h>
#include <math.h>
#include "nr.h"
#include "nrutil.h"

#define JMAX 11
#define PI 3.1415926

main()
{
    int i,j,midl;
    double **a,**b,**c,**d,**e,**f,**u,rjac;

    a=dmatrix(1,JMAX,1,JMAX);
    b=dmatrix(1,JMAX,1,JMAX);
    c=dmatrix(1,JMAX,1,JMAX);
    d=dmatrix(1,JMAX,1,JMAX);
    e=dmatrix(1,JMAX,1,JMAX);
    f=dmatrix(1,JMAX,1,JMAX);
    u=dmatrix(1,JMAX,1,JMAX);
    for (i=1;i<=JMAX;i++)
        for (j=1;j<=JMAX;j++) {
            a[i][j]=b[i][j]=c[i][j]=d[i][j]=1.0;
            e[i][j]=(-4.0);
            f[i][j]=u[i][j]=0.0;
```

```
        }
    midl=JMAX/2+1;
    f[midl][midl]=2.0;
    rjac=cos(PI/JMAX);
    sor(a,b,c,d,e,f,u,JMAX,rjac);
    printf("SOR solution:\n");
    for (i=1;i<=JMAX;i++) {
        for (j=1;j<=JMAX;j++) printf("%7.2f",u[i][j]);
        printf("\n");
    }
    printf("\n Test that solution satisfies difference equations:\n");
    for (i=2;i<JMAX;i++) {
        for (j=2;j<JMAX;j++)
            f[i][j]=u[i+1][j]+u[i-1][j]+u[i][j+1]+u[i][j-1]
                -4.0*u[i][j];
        printf("%7s"," ");
        for (j=2;j<JMAX;j++) printf("%7.2f",f[i][j]);
        printf("\n");
    }
    free_dmatrix(u,1,JMAX,1,JMAX);
    free_dmatrix(f,1,JMAX,1,JMAX);
    free_dmatrix(e,1,JMAX,1,JMAX);
    free_dmatrix(d,1,JMAX,1,JMAX);
    free_dmatrix(c,1,JMAX,1,JMAX);
    free_dmatrix(b,1,JMAX,1,JMAX);
    free_dmatrix(a,1,JMAX,1,JMAX);
}
```

Routine `adi` uses the alternating direction implicit method for solving partial differential equations. This method can be considerably more efficient than the `sor` calculation, and is preferred among relaxation methods when the shape of the grid and the boundary conditions allow its use. It is admittedly slightly more difficult to program, and sometimes does not converge, but it is the recommended "first-try" algorithm. Sample program `xadi.c` uses the same model problem outlined above. When it is subjected to operator splitting and put in the form of equations (17.6.22) of *Numerical Recipes*, the coefficient arrays become

$$A_{jl} = C_{jl} = D_{jl} = F_{jl} = -1.0$$

$$B_{jl} = E_{jl} = 2.0$$

Again the trial solution is set to zero everywhere, and the source term is zeroed except at the centerpoint of the grid. As given in the text (equation 17.6.20) bounds on the eigenvalues are

$$\texttt{alpha} = 2\left[1 - \cos\left(\frac{\pi}{\text{JMAX}}\right)\right]$$

$$\texttt{beta} = 2\left[1 - \cos\frac{(\text{JMAX} - 1)\pi}{\text{JMAX}}\right]$$

where $\text{JMAX} \times \text{JMAX}$ is the dimension of the grid. The number of iterations 2^k is minimized by choosing it to be about $\ln(4\text{JMAX}/\pi)$. As in routine `sor`, the solution for u is printed out and may be compared with the copy listed before program `xsor.c`. Also, this solution is substituted into the difference equation and should give a zero result everywhere except at the centerpoint of the grid, where its value is 2.0. Notice that `adi` makes calls to `tridag` and requires a double precision version of that routine, if available.

```c
/* Driver for routine ADI */

#include <stdio.h>
#include <math.h>
#include "nr.h"
#include "nrutil.h"

#define JMAX 11
#define PI 3.1415926

main()
{
    int i,j,k,mid,twotok;
    double alim,alpha,beta,eps;
    double **a,**b,**c,**d,**e,**f,**g,**u;

    a=dmatrix(1,JMAX,1,JMAX);
    b=dmatrix(1,JMAX,1,JMAX);
    c=dmatrix(1,JMAX,1,JMAX);
    d=dmatrix(1,JMAX,1,JMAX);
    e=dmatrix(1,JMAX,1,JMAX);
    f=dmatrix(1,JMAX,1,JMAX);
    g=dmatrix(1,JMAX,1,JMAX);
    u=dmatrix(1,JMAX,1,JMAX);
    for (i=1;i<=JMAX;i++) {
        for (j=1;j<=JMAX;j++) {
            a[i][j] = -1.0;
            b[i][j]=2.0;
            c[i][j] = -1.0;
            d[i][j] = -1.0;
            e[i][j]=2.0;
            f[i][j] = -1.0;
            g[i][j]=0.0;
            u[i][j]=0.0;
        }
    }
    mid=JMAX/2+1;
    g[mid][mid]=2.0;
    alpha=2.0*(1.0-cos(PI/JMAX));
    beta=2.0*(1.0-cos((JMAX-1)*PI/JMAX));
    alim=log(4.0*JMAX/PI);
    k=0;
    twotok=1;
    while (twotok < alim) {
        ++k;
        twotok *= 2;
    }
    eps=1.0e-4;
    adi(a,b,c,d,e,f,g,u,JMAX,k,alpha,beta,eps);
    printf("ADI Solution:\n");
    for (i=1;i<=JMAX;i++) {
        for (j=1;j<=JMAX;j++) printf("%7.2f",u[i][j]);
        printf("\n");
    }
```

```
    printf("\nTest that solution satisfies difference eqns:\n");
    for (i=2;i<=JMAX-1;i++) {
        for (j=2;j<=JMAX-1;j++)
            g[i][j] = -4.0*u[i][j]+u[i+1][j]
                +u[i-1][j]+u[i][j-1]+u[i][j+1];
        printf("          ");
        for (j=2;j<=JMAX-1;j++) printf("%7.2f",g[i][j]);
        printf("\n");
    }
    free_dmatrix(u,1,JMAX,1,JMAX);
    free_dmatrix(g,1,JMAX,1,JMAX);
    free_dmatrix(f,1,JMAX,1,JMAX);
    free_dmatrix(e,1,JMAX,1,JMAX);
    free_dmatrix(d,1,JMAX,1,JMAX);
    free_dmatrix(c,1,JMAX,1,JMAX);
    free_dmatrix(b,1,JMAX,1,JMAX);
    free_dmatrix(a,1,JMAX,1,JMAX);
}
```

Appendix A: Header Files

The Numerical Recipes functions, and the example routines in this book, make use of the following header files. The files listed here are abbreviated forms of those found on the Numerical Recipes diskettes, in that they do not show the alternative constructions for use with compilers that do not support the full prototype specifications of the ANSI C standard. However, the prototypes listed here may be shortened for use with older compilers, and they are a helpful in checking the data types expected by each function.

$$\star \quad \star \quad \star \quad \star$$

File nrutil.h contains a number of utility routines, which are discussed in Chapter 1 of *Numerical Recipes: The Art of Scientific Computing (C)*. They are used for error reporting, dynamic allocation and deallocation of memory for vectors and matrices, and the creation of references to submatrices. Source code for the utility functions is listed in Appendix B.

```
#ifdef ANSI
    void nrerror(char *error_text);
    float *vector(int nl, int nh);
    int *ivector(int nl,int nh);
    double *dvector(int nl, int nh);
    float **matrix(int nrl, int nrh, int ncl, int nch);
    int **imatrix(int nrl, int nrh, int ncl, int nch);
    double **dmatrix(int nrl, int nrh, int ncl, int nch);
    float **submatrix(float **a, int oldrl, int oldrh, int oldcl,
        int oldch, int newrl, int newcl);
    float **convert_matrix(float *a, int nrl, int nrh, int ncl, int nch);
    void free_vector(float *v, int nl, int nh);
    void free_ivector(int *v, int nl, int nh);
    void free_dvector(double *v, int nl, int nh);
    void free_matrix(float **m, int nrl,int nrh, int ncl, int nch);
    void free_imatrix(int **m, int nrl, int nrh, int ncl, int nch);
    void free_dmatrix(double **m, int nrl, int nrh, int ncl, int nch);
    void free_submatrix(float **b, int nrl, int nrh, int ncl, int nch);
    void free_convert_matrix(float **b, int nrl, int nrh, int ncl, int nch);
#endif
```

The complex.h header file contains prototypes for a set of routines which perform arithmetic functions with complex variables. Complex data types, and predefined arithmetic operations on complex numbers, are not part of the ANSI C standard. However, if your C library provides such facilities, you may use them in place of the ones we provide. Source code for these functions may be found in Appendix C.

```
typedef struct FCOMPLEX {float r,i;} fcomplex;

#ifdef ANSI
    fcomplex Cadd(fcomplex a, fcomplex b);
```

223

```
    fcomplex Csub(fcomplex a, fcomplex b);
    fcomplex Cmul(fcomplex a, fcomplex b);
    fcomplex Complex(float re, float im);
    fcomplex Conjg(fcomplex z);
    fcomplex Cdiv(fcomplex a, fcomplex b);
    float Cabs(fcomplex z);
    fcomplex Csqrt(fcomplex z);
    fcomplex RCmul(float x, fcomplex a);
#endif
```

The following is a list of prototypes for all functions in the Numerical Recipes software collection. This is a section of the file nr.h, a header file included in virtually all of the example routines in this book. As an alternative to including nr.h in your programs, you may cull from this file the references to the specific Recipes which you will be using, and incorporate them as individual declarations.

```
    typedef struct FCOMPLEX {float r,i;} fcomplex;
    typedef struct IMMENSE {unsigned long l,r;} immense;
    typedef struct GREAT {unsigned short l,c,r;} great;

#ifdef ANSI
    void   adi(double **a, double **b, double **c, double **d, double **e,
                double **f, double **g, double **u, int jmax, int k,
                double alpha, double beta, double eps);
    void   amoeba(float **p, float *y, int ndim, float ftol,
                float (*funk)(float *), int *iter);
    void   anneal(float *x, float *y, int *iorder, int ncity);
    void   avevar(float *data, int n, float *ave, float *svar);
    void   balanc(float **a, int n);
    void   bcucof(float *y, float *y1, float *y2, float *y12, float d1,
                float d2, float **c);
    void   bcuint(float *y, float *y1, float *y2, float *y12, float x11,
                float x1u, float x21, float x2u, float x1, float x2,
                float *ansy, float *ansy1, float *ansy2);
    float  bessi(int n, float x);
    float  bessi0(float x);
    float  bessi1(float x);
    float  bessj(int n, float x);
    float  bessj0(float x);
    float  bessj1(float x);
    float  bessk(int n, float x);
    float  bessk0(float x);
    float  bessk1(float x);
    float  bessy(int n, float x);
    float  bessy0(float x);
    float  bessy1(float x);
    float  beta(float z, float w);
    float  betacf(float a, float b, float x);
    float  betai(float a, float b, float x);
    float  bico(int n, int k);
    void   bksub(int ne, int nb, int jf, int k1, int k2, float ***c);
    float  bnldev(float pp, int n, int *idum);
    float  brent(float ax, float bx, float cx, float (*f)(float), float tol,
                float *xmin);
    void   bsstep(float *y, float *dydx, int nv, float *xx, float htry,
                float eps, float *yscal, float *hdid, float *hnext,
                void (*derivs)(float,float *,float *));
    void   caldat(long julian, int *mm, int *id, int *iyyy);
    float  cel(float qqc, float pp, float aa, float bb);
    void   chder(float a, float b, float *c, float *cder, int n);
    float  chebev(float a, float b, float *c, int m, float x);
    void   chebft(float a, float b, float *c, int n, float (*func)(float));
```

```
void  chebpc(float *c, float *d, int n);
void  chint(float a, float b, float *c, float *cint, int n);
void  chsone(float *bins, float *ebins, int nbins, int knstrn,
             float *df, float *chsq, float *prob);
void  chstwo(float *bins1, float *bins2, int nbins, int knstrn,
             float *df, float *chsq, float *prob);
void  cntab1(int **nn, int ni, int nj, float *chisq, float *df,
             float *prob, float *cramrv, float *ccc);
void  cntab2(int **nn, int ni, int nj, float *h, float *hx, float *hy,
             float *hygx, float *hxgy, float *uygx, float *uxgy,
             float *uxy);
void  convlv(float *data, int n, float *respns, int m, int isign,
             float *ans);
void  correl(float *data1, float *data2, int n, float *ans);
void  cosft(float *y, int n, int isign);
void  covsrt(float **covar, int ma, int *lista, int mfit);
void  crank(int n, float *w, float *s);
float dbrent(float ax, float bx, float cx, float (*f)(float),
             float (*df)(float), float tol, float *xmin);
void  ddpoly(float *c, int nc, float x, float *pd, int nd);
void  des(immense inp, immense key, int *newkey, int isw, immense *out);
void  ks(immense key, int n, great *kn);
void  cyfun(unsigned long ir, great k, unsigned long *iout);
float df1dim(float x);
void  dfpmin(float *p, int n, float ftol, int *iter, float *fret,
             float (*func)(float *), void (*dfunc)(float *,float *));
void  difeq(int k, int k1, int k2, int jsf, int is1, int isf,
            int *indexv, int ne, float **s, float **y);
void  dlinmin(float *p, float *xi, int n, float *fret,
             float (*func)(float *), void (*dfunc)(float *,float *));
void  eclass(int *nf, int n, int *lista, int *listb, int m);
void  eclazz(int *nf, int n, int (*equiv)(int,int));
void  eigsrt(float *d, float **v, int n);
float el2(float x, float qqc, float aa, float bb);
void  elmhes(float **a, int n);
float erf(float x);
float erfc(float x);
float erfcc(float x);
void  eulsum(float *sum, float term, int jterm, float *wksp);
float evlmem(float fdt, float *cof, int m, float pm);
float expdev(int *idum);
float f1dim(float x);
float factln(int n);
float factrl(int n);
void  fgauss(float x, float *a, float *y, float *dyda, int na);
void  fit(float *x, float *y, int ndata, float *sig, int mwt, float *a,
          float *b, float *siga, float *sigb, float *chi2, float *q);
void  fixrts(float *d, int npoles);
void  fleg(float x, float *pl, int nl);
void  flmoon(int n, int nph, long *jd, float *frac);
void  four1(float *data, int nn, int isign);
void  fourn(float *data, int *nn, int ndim, int isign);
void  fpoly(float x, float *p, int np);
void  frprmn(float *p, int n, float ftol, int *iter, float *fret,
             float (*func)(float *), void (*dfunc)(float *,float *));
void  ftest(float *data1, int n1, float *data2, int n2, float *f,
            float *prob);
float gamdev(int ia, int *idum);
float gammln(float xx);
float gammp(float a, float x);
float gammq(float a, float x);
float gasdev(int *idum);
```

```
void   gauleg(double x1, double x2, double *x, double *w, int n);
void   gaussj(float **a, int n, float **b, int m);
void   gcf(float *gammcf, float a, float x, float *gln);
float  golden(float ax, float bx, float cx, float (*f)(float), float tol,
             float *xmin);
void   gser(float *gamser, float a, float x, float *gln);
void   hqr(float **a, int n, float *wr, float *wi);
void   hunt(float *xx, int n, float x, int *jlo);
void   indexx(int n, float *arrin, int *indx);
int    irbit1(unsigned long int *iseed);
int    irbit2(unsigned long int *iseed);
void   jacobi(float **a, int n, float *d, float **v, int *nrot);
long   julday(int mm, int id, int iyyy);
void   kendl1(float *data1, float *data2, int n, float *tau, float *z,
             float *prob);
void   kendl2(float **tab, int i, int j, float *tau, float *z,
             float *prob);
void   ksone(float *data, int n, float (*func)(float), float *d,
             float *prob);
void   kstwo(float *data1, int n1, float *data2, int n2, float *d,
             float *prob);
void   laguer(fcomplex *a, int m, fcomplex *x, float eps, int polish);
void   lfit(float *x, float *y, float *sig, int ndata, float *a, int ma,
             int *lista, int mfit, float **covar, float *chisq,
             void (*funcs)(float,float *,int));
void   linmin(float *p, float *xi, int n, float *fret, float (*func)(float));
void   locate(float *xx, int n, float x, int *j);
void   lubksb(float **a, int n, int *indx, float *b);
void   ludcmp(float **a, int n, int *indx, float *d);
void   mdian1(float *x, int n, float *xmed);
void   mdian2(float *x, int n, float *xmed);
void   medfit(float *x, float *y, int ndata, float *a, float *b,
             float *abdev);
void   memcof(float *data, int n, int m, float *pm, float *cof);
float  midexp(float (*funk)(float), float aa, float bb, int n);
float  midinf(float (*funk)(float), float aa, float bb, int n);
float  midpnt(float (*func)(float), float a, float b, int n);
float  midsql(float (*funk)(float), float aa, float bb, int n);
float  midsqu(float (*funk)(float), float aa, float bb, int n);
void   mmid(float *y, float *dydx, int nvar, float xs, float htot,
             int nstep, float *yout,
             void (*derivs)(float,float *,float *));
void   mnbrak(float *ax, float *bx, float *cx, float *fa, float *fb,
             float *fc, float (*func)(float));
void   mnewt(int ntrial, float *x, int n, float tolx, float tolf);
void   moment(float *data, int n, float *ave, float *adev, float *sdev,
             float *svar, float *skew, float *curt);
void   mprove(float **a, float **alud, int n, int *indx, float *b,
             float *x);
void   mrqcof(float *x, float *y, float *sig, int ndata, float *a, int ma,
             int *lista, int mfit, float **alpha, float *beta, float
             *chisq, void (*funcs)(float,float *,float *,float *,int));
void   mrqmin(float *x, float *y, float *sig, int ndata, float *a,
             int ma, int *lista, int mfit, float **covar, float **alpha,
             float *chisq, void (*funcs)(float,float *,float *,float *,
             int),float *alamda);
void   odeint(float *ystart, int nvar, float x1, float x2, float eps,
             float h1, float hmin, int *nok, int *nbad,
             void (*derivs)(float,float *,float *),
             void (*rkqc)(float *,float *,int,float *,float,float,float
             *,float *,float *,void (*)(float,float *,float *)));
void   pcshft(float a, float b, float *d, int n);
```

```
void   pearsn(float *x, float *y, int n, float *r, float *prob, float *z);
void   piksr2(int n, float *arr, float *brr);
void   piksrt(int n, float *arr);
void   pinvs(int ie1, int ie2, int je1, int jsf, int jc1, int k,
             float ***c, float **s);
float  plgndr(int l, int m, float x);
float  poidev(float xm, int *idum);
void   polcoe(float *x, float *y, int n, float *cof);
void   polcof(float *xa, float *ya, int n, float *cof);
void   poldiv(float *u, int n, float *v, int nv, float *q, float *r);
void   polin2(float *x1a, float *x2a, float **ya, int m, int n, float x1,
             float x2, float *y, float *dy);
void   polint(float *xa, float *ya, int n, float x, float *y, float *dy);
void   powell(float *p, float **xi, int n, float ftol, int *iter,
             float *fret, float (*func)(float *));
void   predic(float *data, int ndata, float *d, int npoles,
             float *future, int nfut);
float  probks(float alam);
void   pzextr(int iest, float xest, float *yest, float *yz, float *dy,
             int nv, int nuse);
void   qcksrt(int n, float *arr);
float  qgaus(float (*func)(float), float a, float b);
float  qromb(float (*func)(float), float a, float b);
float  qromo(float (*func)(float), float a, float b,
             float (*choose)(float (*)(float),float,float,int));
void   qroot(float *p, int n, float *b, float *c, float eps);
float  qsimp(float (*func)(float), float a, float b);
float  qtrap(float (*func)(float), float a, float b);
float  quad3d(float (*func)(float,float,float), float x1, float x2);
float  ran0(int *idum);
float  ran1(int *idum);
float  ran2(long *idum);
float  ran3(int *idum);
float  ran4(int *idum);
void   rank(int n, int *indx, int *irank);
void   ratint(float *xa, float *ya, int n, float x, float *y, float *dy);
void   realft(float *data, int n, int isign);
void   red(int iz1, int iz2, int jz1, int jz2, int jm1, int jm2, int jmf,
             int ic1, int jc1, int jcf, int kc, float ***c, float **s);
void   rk4(float *y, float *dydx, int n, float x, float h, float *yout,
             void (*derivs)(float,float *,float *));
void   rkdumb(float *vstart, int nvar, float x1, float x2, int nstep,
             void (*derivs)(float,float *,float *));
void   rkqc(float *y, float *dydx, int n, float *x, float htry,
             float eps, float *yscal, float *hdid, float *hnext,
             void (*derivs)(float,float *,float *));
float  rofunc(float b);
float  rtbis(float (*func)(float), float x1, float x2, float xacc);
float  rtflsp(float (*func)(float), float x1, float x2, float xacc);
float  rtnewt(void (*funcd)(float,float *,float *), float x1, float x2,
             float xacc);
float  rtsafe(void (*funcd)(float,float *,float *), float x1, float x2,
             float xacc);
float  rtsec(float (*func)(float), float x1, float x2, float xacc);
void   rzextr(int iest, float xest, float *yest, float *yz, float *dy,
             int nv, int nuse);
void   scrsho(float (*fx)(float));
void   shell(int n, float *arr);
void   shoot(int nvar, float *v, float *delv, int n2, float x1, float x2,
             float eps, float h1, float hmin, float *f, float *dv);
void   shootf(int nvar, float *v1, float *v2, float *delv1, float *delv2,
             int n1, int n2, float x1, float x2, float xf, float eps,
```

```
                        float h1, float hmin, float *f, float *dv1, float *dv2);
     void  simp1(float **a, int mm, int *ll, int nll, int iabf, int *kp,
                        float *bmax);
     void  simp2(float **a, int n, int *l2, int nl2, int *ip, int kp,
                        float *q1);
     void  simp3(float **a, int i1, int k1, int ip, int kp);
     void  simplx(float **a, int m, int n, int m1, int m2, int m3,
                        int *icase, int *izrov, int *iposv);
     void  sinft(float *y, int n);
     void  smooft(float *y, int n, float pts);
     void  sncndn(float uu, float emmc, float *sn, float *cn, float *dn);
     void  solvde(int itmax, float conv, float slowc, float *scalv,
                        int *indexv, int ne, int nb, int m, float **y, float ***c,
                        float **s);
     void  sor(double **a, double **b, double **c, double **d, double **e,
                        double **f, double **u, int jmax, double rjac);
     void  sort(int n, float *ra);
     void  sort2(int n, float *ra, float *rb);
     void  sort3(int n, float *ra, float *rb, float *rc);
     void  sparse(float *b, int n, float *x, float *rsq);
     void  spctrm(FILE *fp, float *p, int m, int k, int ovrlap);
     void  spear(float *data1, float *data2, int n, float *d, float *zd,
                        float *probd, float *rs, float *probrs);
     void  splie2(float *x1a, float *x2a, float **ya, int m, int n,
                        float **y2a);
     void  splin2(float *x1a, float *x2a, float **ya, float **y2a, int m,
                        int n, float x1, float x2, float *y);
     void  spline(float *x, float *y, int n, float yp1, float ypn, float *y2);
     void  splint(float *xa, float *ya, float *y2a, int n, float x, float *y);
     void  svbksb(float **u, float *w, float **v, int m, int n, float *b,
                        float *x);
     void  svdcmp(float **a, int m, int n, float *w, float **v);
     void  svdfit(float *x, float *y, float *sig, int ndata, float *a,
                        int ma, float **u, float **v, float *w, float *chisq,
                        void (*funcs)(float,float *,int));
     void  svdvar(float **v, int ma, float *w, float **cvm);
     void  toeplz(float *r, float *x, float *y, int n);
     void  tptest(float *data1, float *data2, int n, float *t, float *prob);
     void  tqli(float *d, float *e, int n, float **z);
     float trapzd(float (*func)(float), float a, float b, int n);
     void  tred2(float **a, int n, float *d, float *e);
     void  tridag(float *a, float *b, float *c, float *r, float *u, int n);
     void  ttest(float *data1, int n1, float *data2, int n2, float *t,
                        float *prob);
     void  tutest(float *data1, int n1, float *data2, int n2, float *t,
                        float *prob);
     void  twofft(float *data1, float *data2, float *fft1, float *fft2,
                        int n);
     void  vander(float *x, float *w, float *q, int n);
     int   zbrac(float (*func)(float), float *x1, float *x2);
     void  zbrak(float (*fx)(float), float x1, float x2, int n, float *xb1,
                        float *xb2, int *nb);
     float zbrent(float (*func)(float), float x1, float x2, float tol);
     void  zroots(fcomplex *a, int m, fcomplex *roots, int polish);
#endif
```

Appendix B: Numerical Recipes Utility Functions

The utility functions listed below are used by many Recipes and Examples in the Numerical Recipes collection. Along with the short error-reporting function, there are functions that we have found indispensible in the handling of vectors and matrices of different data types. A full discussion of the use of matrices and vectors within the Numerical Recipes collection may be found in Chapter 1 of Numerical Recipes in C: The Art of Scientific Computing, *along with a description of other programming conventions adopted for the use of C in scientific programming.*

★   ★   ★   ★

```c
#include <malloc.h>
#include <stdio.h>

void nrerror(error_text)
char error_text[];
{
    void exit();

    fprintf(stderr,"Numerical Recipes run-time error...\n");
    fprintf(stderr,"%s\n",error_text);
    fprintf(stderr,"...now exiting to system...\n");
    exit(1);
}



float *vector(nl,nh)
int nl,nh;
{
    float *v;

    v=(float *)malloc((unsigned) (nh-nl+1)*sizeof(float));
    if (!v) nrerror("allocation failure in vector()");
    return v-nl;
}

int *ivector(nl,nh)
int nl,nh;
{
    int *v;

    v=(int *)malloc((unsigned) (nh-nl+1)*sizeof(int));
    if (!v) nrerror("allocation failure in ivector()");
```

```
        return v-nl;
}

double *dvector(nl,nh)
int nl,nh;
{
    double *v;

    v=(double *)malloc((unsigned) (nh-nl+1)*sizeof(double));
    if (!v) nrerror("allocation failure in dvector()");
    return v-nl;
}




float **matrix(nrl,nrh,ncl,nch)
int nrl,nrh,ncl,nch;
{
    int i;
    float **m;

    m=(float **) malloc((unsigned) (nrh-nrl+1)*sizeof(float*));
    if (!m) nrerror("allocation failure 1 in matrix()");
    m -= nrl;

    for(i=nrl;i<=nrh;i++) {
        m[i]=(float *) malloc((unsigned) (nch-ncl+1)*sizeof(float));
        if (!m[i]) nrerror("allocation failure 2 in matrix()");
        m[i] -= ncl;
    }
    return m;
}

double **dmatrix(nrl,nrh,ncl,nch)
int nrl,nrh,ncl,nch;
{
    int i;
    double **m;

    m=(double **) malloc((unsigned) (nrh-nrl+1)*sizeof(double*));
    if (!m) nrerror("allocation failure 1 in dmatrix()");
    m -= nrl;

    for(i=nrl;i<=nrh;i++) {
        m[i]=(double *) malloc((unsigned) (nch-ncl+1)*sizeof(double));
        if (!m[i]) nrerror("allocation failure 2 in dmatrix()");
        m[i] -= ncl;
    }
    return m;
}

int **imatrix(nrl,nrh,ncl,nch)
int nrl,nrh,ncl,nch;
{
    int i,**m;

    m=(int **)malloc((unsigned) (nrh-nrl+1)*sizeof(int*));
```

```
    if (!m) nrerror("allocation failure 1 in imatrix()");
    m -= nrl;

    for(i=nrl;i<=nrh;i++) {
        m[i]=(int *)malloc((unsigned) (nch-ncl+1)*sizeof(int));
        if (!m[i]) nrerror("allocation failure 2 in imatrix()");
        m[i] -= ncl;
    }
    return m;
}



float **submatrix(a,oldrl,oldrh,oldcl,oldch,newrl,newcl)
float **a;
int oldrl,oldrh,oldcl,oldch,newrl,newcl;
{
    int i,j;
    float **m;

    m=(float **) malloc((unsigned) (oldrh-oldrl+1)*sizeof(float*));
    if (!m) nrerror("allocation failure in submatrix()");
    m -= newrl;

    for(i=oldrl,j=newrl;i<=oldrh;i++,j++) m[j]=a[i]+oldcl-newcl;

    return m;
}



void free_vector(v,nl,nh)
float *v;
int nl,nh;
{
    free((char*) (v+nl));
}

void free_ivector(v,nl,nh)
int *v,nl,nh;
{
    free((char*) (v+nl));
}

void free_dvector(v,nl,nh)
double *v;
int nl,nh;
{
    free((char*) (v+nl));
}



void free_matrix(m,nrl,nrh,ncl,nch)
float **m;
int nrl,nrh,ncl,nch;
{
```

```
    int i;

    for(i=nrh;i>=nrl;i--) free((char*) (m[i]+ncl));
    free((char*) (m+nrl));
}

void free_dmatrix(m,nrl,nrh,ncl,nch)
double **m;
int nrl,nrh,ncl,nch;
{
    int i;

    for(i=nrh;i>=nrl;i--) free((char*) (m[i]+ncl));
    free((char*) (m+nrl));
}

void free_imatrix(m,nrl,nrh,ncl,nch)
int **m;
int nrl,nrh,ncl,nch;
{
    int i;

    for(i=nrh;i>=nrl;i--) free((char*) (m[i]+ncl));
    free((char*) (m+nrl));
}



void free_submatrix(b,nrl,nrh,ncl,nch)
float **b;
int nrl,nrh,ncl,nch;
{
    free((char*) (b+nrl));
}



float **convert_matrix(a,nrl,nrh,ncl,nch)
float *a;
int nrl,nrh,ncl,nch;
{
    int i,j,nrow,ncol;
    float **m;

    nrow=nrh-nrl+1;
    ncol=nch-ncl+1;
    m = (float **) malloc((unsigned) (nrow)*sizeof(float*));
    if (!m) nrerror("allocation failure in convert_matrix()");
    m -= nrl;
    for(i=0,j=nrl;i<=nrow-1;i++,j++) m[j]=a+ncol*i-ncl;
    return m;
}



void free_convert_matrix(b,nrl,nrh,ncl,nch)
float **b;
```

```
int nrl,nrh,ncl,nch;
{
    free((char*) (b+nrl));
}
```

Appendix C: Functions for Complex Arithmetic

Complex data types, and arithmetic operations on complex numbers, are not a standard part of C. We have therefore included the following functions for use with the small number of Numerical Recipes routines which use complex variables. When using C libraries which support the use of complex numbers, references to these routines may be replaced by equivalent library functions.

★ ★ ★ ★

```c
#include <math.h>

typedef struct FCOMPLEX {float r,i;} fcomplex;

fcomplex Cadd(a,b)
fcomplex a,b;
{    fcomplex c;
    c.r=a.r+b.r;
    c.i=a.i+b.i;
    return c;
}

fcomplex Csub(a,b)
fcomplex a,b;
{    fcomplex c;
    c.r=a.r-b.r;
    c.i=a.i-b.i;
    return c;
}

fcomplex Cmul(a,b)
fcomplex a,b;
{    fcomplex c;
    c.r=a.r*b.r-a.i*b.i;
    c.i=a.i*b.r+a.r*b.i;
    return c;
}

fcomplex Complex(re,im)
float re,im;
{    fcomplex c;
    c.r=re;
    c.i=im;
    return c;
}
```

234

```
fcomplex Conjg(z)
fcomplex z;
{    fcomplex c;
    c.r=z.r;
    c.i = -z.i;
    return c;
}

fcomplex Cdiv(a,b)
fcomplex a,b;
{    fcomplex c;
    float r,den;
    if (fabs(b.r) >= fabs(b.i)) {
        r=b.i/b.r;
        den=b.r+r*b.i;
        c.r=(a.r+r*a.i)/den;
        c.i=(a.i-r*a.r)/den;
    } else {
        r=b.r/b.i;
        den=b.i+r*b.r;
        c.r=(a.r*r+a.i)/den;
        c.i=(a.i*r-a.r)/den;
    }
    return c;
}

float Cabs(z)
fcomplex z;
{    float x,y,ans,temp;
    x=fabs(z.r);
    y=fabs(z.i);
    if (x == 0.0)
        ans=y;
    else if (y == 0.0)
        ans=x;
    else if (x > y) {
        temp=y/x;
        ans=x*sqrt(1.0+temp*temp);
    } else {
        temp=x/y;
        ans=y*sqrt(1.0+temp*temp);
    }
    return ans;
}

fcomplex Csqrt(z)
fcomplex z;
{    fcomplex c;
    float x,y,w,r;
    if ((z.r == 0.0) && (z.i == 0.0)) {
        c.r=0.0;
        c.i=0.0;
        return c;
    } else {
        x=fabs(z.r);
        y=fabs(z.i);
        if (x >= y) {
```

```
            r=y/x;
            w=sqrt(x)*sqrt(0.5*(1.0+sqrt(1.0+r*r)));
        } else {
            r=x/y;
            w=sqrt(y)*sqrt(0.5*(r+sqrt(1.0+r*r)));
        }
        if (z.r >= 0.0) {
            c.r=w;
            c.i=z.i/(2.0*w);
        } else {
            c.i=(z.i >= 0) ? w : -w;
            c.r=z.i/(2.0*c.i);
        }
        return c;
    }
}

fcomplex RCmul(x,a)
float x;
fcomplex a;
{   fcomplex c;
    c.r=x*a.r;
    c.i=x*a.i;
    return c;
}
```

Index of Demonstrated Procedures

adi	218, 220		cntab2	165, 177
amoeba	125, 129		convlv	150, 157
anneal	125, 138		correl	150, 158
asub	17		cosft	150, 156
atsub	17		covsrt	187, 190
avevar	91, 168		crank	180
badluk	1, 3		cyfun	81, 94
balanc	140, 146, 147, 148		dbrent	125, 127
bcucof	22, 31, 32		ddpoly	44, 45
bcuint	22, 31, 32		des	81, 91, 94
bessi	68		desks	94
bessi0	65		df1dim	135
bessi1	66		dfpmin	125, 136
bessj	64, 205		dlinmin	134
bessj0	61		eclass	102, 110
bessj1	63, 127		eclazz	102, 111
bessk	67		eigsrt	140, 142
bessk0	65		el2	69
bessk1	67		elmhes	140, 147, 148
bessy	64		erf	59, 173
bessy0	62		erfc	60
bessy1	63		erfcc	60
beta	56		eulsum	44
betai	61		evlmem	150, 159, 160
betacf	61		expdev	81, 85, 172
bico	55		factln	55
bnldev	81, 87		factrl	54
brent	125, 127		f1dim	132, 135
bsstep	202, 208, 209		fgauss	195, 198
caldat	1, 3		fit	187, 199
cel	69		fixrts	150, 161
chder	44, 49		fleg	194
chebev	44, 48, 49		flmoon	1
chebft	44, 47, 49, 51		four1	150, 152
chebpc	44, 51, 52		fourn	150, 163
chint	44, 49		fpoly	194
chisq	195		frprmn	125, 133, 136
chsone	165, 172		ftest	165, 171
chstwo	165, 172		gamdev	81, 87
cntab1	165, 176		gammln	53

gammp	57	piksrt	102, 104, 105, 109
gammq	57	plgndr	68, 194
gasdev	81, 86, 166, 167, 169	poidev	81, 87
gauleg	36, 41	polcoe	22, 28
gaussj	5, 9	polcof	22, 28
gcf	58	poldiv	44, 46
golden	125, 126, 127	polin2	22, 30
gser	58	polint	22, 23, 39
hqr	140, 148	powell	125, 130
hunt	22, 27	predic	150, 161, 163
indexx	102, 106, 107, 108	probks	175
irbit1	81, 89, 183	pzextr	202, 210
irbit2	81, 89, 183	qcksrt	102, 109
jacobi	140, 142, 144	qgaus	36, 41
julday	2, 3	qromb	36, 39
kendl1	165, 182	qromo	36
kendl2	165, 182, 183	qroot	113, 121
ks	81, 94	qsimp	36, 38, 39, 69
ksone	165, 173, 175	qtrap	36, 37, 39
kstwo	165, 173, 174, 175	quad3d	42
laguer	113, 119, 120, 121	ran0	81, 91, 182
lfit	187, 188, 190, 192	ran1	81
linmin	125, 130, 131, 132, 133, 134	ran2	81
locate	22, 26, 27	ran3	81, 91, 138, 182
lubksb	5, 7, 9, 11, 14	ran4	81, 91, 182
ludcmp	5, 7, 9, 11, 14	rank	102, 108
mdian1	165, 166	ratint	22, 23
mdian2	165, 166	realft	150, 153, 156
medfit	187, 199, 200	rk4	202, 203, 204, 205
memcof	150, 159, 160, 161	rkdumb	202, 203, 205
midexp	36	rkqc	202, 204, 205, 208
midinf	36	rofunc	187, 200
midpnt	36, 39, 40	rtbis	113, 115
midsql	36, 40	rtflsp	113
midsqu	36	rtnewt	113, 116
mmid	202, 206	rtsafe	113, 116
mnbrak	125, 126, 127	rtsec	113
mnewt	113, 122	rzextr	202, 208, 209, 210
moment	165	scrsho	113, 132
mprove	11	sfroid	212
mrqcof	187, 195, 197	shell	102, 104
mrqfit	198	shoot	212, 213
mrqmin	187, 195	shootf	212
odeint	202, 205, 208, 212	simplx	125, 137
pcshft	44, 51, 52	sinft	150, 154, 156
pearsn	165, 178	smooft	165, 184
piksr2	102, 103, 105	sncndn	71

```
solvde     212
sor        218, 219, 220
sort       102, 105, 109
sort2      102, 105, 180
sort3      102, 107
sparse     5, 17
spctrm     150, 159
spear      165, 179, 180
splie2     22, 33, 34
splin2     22, 33, 34
spline     22, 24, 25
splint     22, 25
svbksb     5, 14
svdcmp     5, 14, 15
svdfit     187, 190, 192, 193
svdvar     187, 193
toeplz     5, 13
tptest     165, 169
tqli       140, 144
trapzd     36, 37, 39
tred2      140, 143, 144
tridag     5, 10, 220
ttest      165, 167, 168, 169
tutest     165, 169
twofft     150, 152
vander     5, 12
zbrac      113, 114
zbrak      113, 114
zbrent     113, 116
zroots     113, 120
```